Landslides

*Edited by Yuanzhi Zhang
and Qiuming Cheng*

Published in London, United Kingdom

IntechOpen

Supporting open minds since 2005

Landslides
http://dx.doi.org/10.5772/intechopen.95641
Edited by Yuanzhi Zhang and Qiuming Cheng

Contributors
Md. Azizul Moqsud, Azemeraw Wubalem, Mal Heron, Halil Akinci, Mustafa Zeybek, Sedat Dogan, Denis
Nseka, Vincent Kakembo, Frank Mugagga, Henry Semakula, Hannington Wasswa, Hosea Opedes, Patience
Ayesiga, Seda Çellek, Getnet Mewa, Filagot Mengistu, Fentahun Ayalneh Mekonnen, Adi Susilo, Eko Andi
Suryo, Turniningtyas Rachmawati, Muwardi Sutasoma, Sunaryo Sunaryo, Yıldırım İsmail Tosun, Daniel
Germain, Antonio Jose Teixera Guerra, Sebastien Roy, Peifeng Ma, Yifei Cui, Weixi Wang, Hui Lin,
Yuanzhi Zhang, Yi Zheng, Ka Po Wong, Qiuming Cheng

Notice
Statements and opinions expressed in the chapters are these of the individual contributors and not
necessarily those of the editors or publisher. No responsibility is accepted for the accuracy of
information contained in the published chapters. The publisher assumes no responsibility for any
damage or injury to persons or property arising out of the use of any materials, instructions, methods
or ideas contained in the book.

First published in London, United Kingdom, 2022 by IntechOpen
IntechOpen is the global imprint of INTECHOPEN LIMITED, registered in England and Wales,
registration number: 11086078, 5 Princes Gate Court, London, SW7 2QJ, United Kingdom
Printed in Croatia

British Library Cataloguing-in-Publication Data
A catalogue record for this book is available from the British Library

Additional hard and PDF copies can be obtained from orders@intechopen.com

Landslides
Edited by Yuanzhi Zhang and Qiuming Cheng
p. cm.
Print ISBN 978-1-83969-023-5
Online ISBN 978-1-83969-024-2
eBook (PDF) ISBN 978-1-83969-025-9

We are IntechOpen,
the world's leading publisher of
Open Access books
Built by scientists, for scientists

6,000+
Open access books available

146,000+
International authors and editors

185M+
Downloads

Our authors are among the

156
Countries delivered to

Top 1%
most cited scientists

12.2%
Contributors from top 500 universities

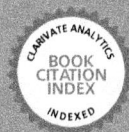

Interested in publishing with us?
Contact book.department@intechopen.com

Meet the editors

Dr. Yuanzhi Zhang is a Research Fellow and Professor of Earth System Science and Coastal Remote Sensing at the Chinese University of Hong Kong and Nanjing University of Information Science and Technology, China. Dr. Zhang received a DSc in Technology from Helsinki University of Technology (now Aalto University), Finland. Dr. Zhang is the author or co-author of 150 peer-reviewed journal articles and 15 books and book chapters. He received the First-Rank Award of the Guangdong Provincial Prize of Science and Technology, China, in 2013, and the Second-Rank Award, Actions for Raising Critical Awareness (ARCA) Prize at the International Symposium "Environment 2010: Situation and Perspectives for the European Union," Porto, Portugal, in 2003. Dr. Zhang was an associate editor for the *International Journal of Applied Earth Observation and Geoinformation*.

Dr. Qiuming Cheng is currently a professor at the School of Earth Science and Engineering, Sun Yat-Sen University, Zhuhai, China, and the founding director of the State Key Lab of Geological Processes and Mineral Resources, China University of Geosciences, Beijing. Dr. Cheng received his Ph.D. in Earth Sciences from the University of Ottawa, Canada, in 1994. He became a faculty member in 1995 and was promoted to full professor in 2002 at York University, Toronto, Canada. Dr. Cheng has published more than 250 refereed papers and 17 book chapters. He received several prestigious awards including the William Christian Krumbein Medal, the highest award given by the International Association for Mathematical Geosciences (IAMG), the AAG Gold Medal, the highest award bestowed by the International Association of Applied Geochemists (AAG), the Canada Foundation for Innovation Award (CFI), and the State Science and Technology Award of the Chinese Government. He served as president of IAMG and is the past president of the International Union of Geological Sciences (IUGS). He is a member of the Chinese Academy of Sciences (CAS) and a foreign member of Academia Europaea (AE).

Contents

Preface

This book presents detailed discussions and documents about landslides as catastrophic events that can cause human injury, loss of life, and economic devastation as well as destroy construction works and cultural and natural heritage. Most of the work in this book is based on in situ observations or remotely sensed data and new technologies in the monitoring of landslide processes, such as coal mine and marine landslides, land subsidence, tsunamis, hazard mapping and assessment, monitoring and modeling, InSAR and GIS techniques, early warning and evacuation, and regional/local landslide mitigation. This book is a comprehensive reference for landslide researchers, graduate students, policymakers, and landslide mitigation managers.

We thank all the contributing authors for their excellent work. We also wish to thank the staff at IntechOpen, especially Publishing Process Manager Ms. Paula Gavran. We acknowledge the use of in situ observations and satellite data from projects at the National Natural Science Foundation of China (U1901215) and the Marine Special Program of Jiangsu Province in China (JSZRHYKJ202007).

Dr. Yuanzhi Zhang
School of Marine Sciences,
Nanjing University of Information Science and Technology,
Nanjing, China

Dr. Qiuming Cheng
School of Earth Sciences and Engineering,
Sun Yat-sen University,
Zhuhai, China

Section 1

Introduction to Landslides

Chapter 1

Introductory Chapter: Landslides

Ka Po Wong, Yuanzhi Zhang and Qiuming Cheng

1. Introduction

1.1 Concept of landslides

Landslides, one of the catastrophic events on earth, can cause extensive impacts, for instance, loss of human life, destruction of infrastructures and residential developments, and damage to cultural and natural heritage [1]. Landslides occur when massive rocks, sand, debris, or a combination of these move downslope, also known as slumps and slope failure. **Table 1** demonstrates the types of landslides movement, namely, falling, toppling, rotational sliding, translational sliding, lateral spreading and flowing [2]. Most landslides occur in mountainous regions; however, the incremental human activities in the natural environment lead to landslides in low-relief areas, such as roadways, river bluff failures, and building excavations. Each occurrence of the landslide has multiple causes (see **Table 1**).

2. Causes of landslides

The leading causes and triggering mechanisms of landslides are physical, natural and human causes [3]. The physical causes include intense rainfall, prolonged intense precipitation, flooding, the rapid drawdown of floods and tides, rapid snowmelt, earthquake, volcanic eruption, thawing, freeze-and-thaw weathering and shrink-and-swell weathering. Natural causes consist of geological and morphological causes. Geological causes include sensitive materials, weathered materials, sheared materials, fissured materials, adversely oriented mass discontinuity and structural discontinuity, and contrast in permeability and contrast in stiffness. Morphological causes consist of tectonic or volcanic uplift, glacial rebound, glacial melt-water outburst, fluvial erosion of slope toe, wave erosion

Type of movement		Type of material		
		Rock	**Debris**	**Earth**
Falls		Rockfall	Debris fall	Earth fall
Topples		Rock topple	Debris topple	Earth topple
Slides	Rotational	Rock slump	Debris slump	Earth slump
	Translational	Rock slide	Debris slide	Earth slide
Spreads		Rock spread	Debris spread	Earth spread
Flows		Rock flow	Debris flow	Earth flow
Complex		Combination of two or more types of landslide movement		

Table 1.
Types of landslide movement.

of slope toe, glacial erosion of slope toe, erosion of lateral margins, subterranean erosion, deposition loading slope or its crest and vegetative removal by forest fire and drought [3]. The human causes include rock, soil and slop excavation, unstable earth fills, loading of slope or its crest, drawdown and filling of reservoirs, deforestation, irrigation, mining waste containment, artificial vibration and water leakage from utilities. Human interventions in the natural environment trigger landslide, leading to the increase in the frequency of landslides and the severity of the damages and casualties [4]. Despite the various types of causes of landslides, the three main causes of the most damaging landslides in the world are slope saturation by water (i.e. severe rainfall), seismic activities with extremely high magnitude in steep landslide-prone areas and volcanic activities [5, 6]. The 1964 Alaska Earthquake triggered widespread landslides, causing massive monetary loss and loss of life [5]. An eruption in Mount St. Helens in 1980 triggered a severe landslide [6]. The sudden lateral shock wave hundreds of miles away made the top of the volcano 1,300 feet away. The shock wave and pyroclastic flow passed through the surrounding landscape, flattening forests, melting snow and ice, and creating massive mudslides [6].

3. Landslides in different types

Landslides that occur in urban areas can destroy infrastructures, for example, roads, residential buildings and public power supplies. Rainfall-induced landslides are ubiquitous in many metropolitan cities [7]. Most landslides caused by rainfall are shallow (i.e. less than a few meters deep), small in size, and moving quickly. Many rainfall-induced landslides turn into mudslides as they move along steep slopes, especially those entering the river, where they may mix with additional water and sediment. Apart from the landslide in the urban area, during the life of the reservoirs, some ancient landslides can be reactivated, and potential new landslide can be triggered in the reservoir areas [8]. The failure of the landslides is affected by the increase in pore water pressure and the decrease in the average effective stress. It is considered as a complex slope instability phenomenon because landslides show obvious kinematics in the failure, post-failure and propagation stages [9]. Landslide can also be caused by volcanic activities in which volcanic gas explosions can be triggered [10]. These landslides can form dams, block rivers and bury roads, bridges and houses. Tsunami, which is the results of submarine earthquakes and collapse of coastal volcanoes, can also be triggered by underwater and coastal landslides, like the 1980 Mount St. Helens eruption. When a fast-moving landslide body enters the water or the water is displaced before, and after the fast-moving underwater landslide, a tsunami may be generated upon impact [11].

4. Remote sensing and GIS applications

Numerous researchers have adopted landslide detection techniques to identify the landside boundaries on the land surface. The conventional techniques for landslide detection are geomorphologic field survey and visual analysis of aerially surveyed images (i.e., orthophotographs) [12]. The purposes of geomorphologic field survey and mapping are to detect and map landslides caused by earthquakes and other specific events, observe the types and characteristics of landslides to improve the visual quality of satellite images or orthographic images and investigate and verify the existing slope evolution inventory map developed using different methods [12]. Due to the reduction in the visualization of slope failure and the limited ability of accurate information of the landslide boundary, there are defects in the field mapping of the landslide [12]. Visual analysis of aerially surveyed images is the oldest remote sensing methodology for detecting landslides. This method is unreliable; however, visual analysis of aerial images have widely been used due to the fact that experienced geomorphologists can effortlessly map and identify landslides on aerial images, trained geomorphologists do not need complex technical skills to obtain aerial photographs, the scale and the size of the aerial survey images allow for an extensive spatial range of terrain with a feasible number of images, and due to the large number of aerial survey photographs from the 1950s, investigators were able to analyze slope failures in the same area. The recent techniques developed for landslide detection are analysis of surface morphology using a high-resolution digital elevation model (HR-DEM) and investigation and analysis of satellite imagery (e.g. panchromatic band images, multiple band images, and radar). Surface morphology using HR-DEM is a sophisticated form of high spatial resolution image which is invisible to the naked eye [13]. HR-DEM is derived from airborne LiDAR data in the south and satellite images in the north. Satellite images of the landslide occurrence were demonstrated by modifying the model and the cover type and electromagnetic radiation from VIS to SWIR of the earth surface [14]. Furthermore, deep learning is the recent trend for landslide investigation using remotely sensed images and can be used for surface classification, transformation detection and object detection [15]. To improve the target of remote sensing-based application, using deep learning methods to achieve the latest results based on computer vision. Geographic Information System (GIS) was used to view different types of information simultaneously since maps and other forms of information are sometimes superimposed on each other using GIS [16]. Different types of information are used for constructing layers in GIS analysis, including the topographic map, terrain map, bedrock map, engineering soil map, forest cover map, aerial photography remote sensing and InSAR imaging [16]. The topographic map is used to indicate the gradient of slopes, configuration of terrain and drainage pattern.

The terrain map is to identify depth, material, terrain configuration, geological processes, the surface of drainage and slope gradient. The bedrock map can identify the types of bedrock, surface and subsurface of the structures and rock age over a topographic map base. The engineering soil map is used to identify the types of surficial material type, drainage and the covers of soils and vegetation. The forest cover map can identify the surface of vegetation, topographic features, surface drainage pattern, and soil drainage character. The identifiable features on aerial photographs can help users identify the type of landslide and reasonably evaluate the overburden features. These, in turn, provide a method for estimating the hazard of landslides on site. Most InSAR devices can penetrate fog and rain and can be used in difficult areas to reach on foot. Two satellite images can be merged to demonstrate the ground displacement for indicating any movement that occurs. Thus, the users can use this means to determine if the hillside moves.

5. Landslide risk assessment and management

To mitigate the damage brought by landslides, governments and related departments corporate to develop landslide risk assessment and management to address the uncertainty of the landslide hazards [17]. Recent landslide risk analysis and assessment provided a systematic and rigorous slope engineering practice and management. The framework of a landslide risk assessment and management consists of the estimation of the risk, the decision of acceptance level and the measures to control the unacceptance level of landslides. The issues required to be addressed are the probability of the occurrence of the landslide, runoff behavior of debris, risk threatening human and property and vulnerability assessment of human and property. Morgan et al. [18] generated the formulas for computing the annual probability of loss for an individual life (Eq. (1)) and property value (Eq. (2)).

$$R(\text{DI}) = P(H) \times P(S|H) \times P(T|S) \times V(L|T) \tag{1}$$

where $R(\text{DI})$ is the annual probability of loss of an individual life; $P(H)$ is the annual probability of the landslide event; $P(S|H)$ is the probability of spatial impact given the event; $P(T|S)$ is the probability of temporal impact given the spatial impact; and $V(L|T)$ is the vulnerability of the individual (probability of loss of life given impact).

$$R(\text{PD}) = P(H) \times P(S|H) \times V(P|S) \times E \tag{2}$$

where $R(\text{PD})$ is the annual loss of property value; $P(H)$ is the annual probability of the landslide event; $P(S|H)$ is the probability of the landslide impacting the property; $V(P|S)$ is the proportion of property value lost; E is the element at risk (e.g. the value of the property).

6. Landslide measures

Several measures have been implemented. Artificial slope upgrading is a significant engineering work to improve and enhance the slope [19]. This practice is based on four main principles, namely removal, reinforcement, retention and replacement. There are four standard practices: fill slopes, soil cut slopes, retaining walls, and rock-cut slopes. For fill slopes, substandard fill slopes usually contain loose filling materials that tend to liquefy when they become saturated and subjected to shear. It needs to install soil nails through filling materials and provide a surface grid to connect the soil nail heads [19]. Existing trees can be preserved during the construction process. The soil nails are embedded in the influential underground stratum to ensure sufficient anchorage to prevent pulling out. Regarding soil cut slopes, trimming the slopes to a gentler profile is the usual method. The main construction activity is the excavation and removal of soil material from the slope [20]. Further, retaining walls for restraining the soil are usually used in steep slopes or the landscapes required to be shaped for engineering or construction works [21]. Gravity walls, pilling walls, cantilever walls and anchored walls are the common types of retaining wall as a landslide solution. The type of wall to be adopted is based on the circumstances, and the critical factors are soil type, slope angle, groundwater, and go on. For rock cut slopes, this construction practice is to upgrade the existing rock cut slopes [22]. The stabilization practices are scaling rocks, buttresses, dentition, rock dowels, rock bolts, rakings drains and mesh netting [22]. Importantly, all these practices are to keep the slope safe and make them look natural so that people can live in a safer and better environment.

7. Summary and future perspective

The research related to the types and causes of landslides have been investigated over time and the findings are similar. More regions are recommended to be assessed to ameliorate the accuracy and generalizability of the previous findings. Furthermore, machine learning and deep learning methods have popularly been applied in the landslide susceptibility mapping and landslide identification. The overall robustness of the results generated from machine learning and deep learning is outstanding. Therefore, adopting machine learning and deep learning in detecting landslides is utterly significant for preventing landslides since human and property losses can be mitigated.

Acknowledgements

This research was partially supported by the National Natural Science Foundation (U1901215), the Marine Special Program of Jiangsu Province in China (JSZRHYKJ202007), the Natural Scientific Foundation of Jiangsu Province (BK20181413), and the State Key Lab Fund for Geological Processes and Mineral Resources (2016).

Conflicts of interest

The authors declare no conflict of interest.

Author details

Ka Po Wong[1], Yuanzhi Zhang[2*] and Qiuming Cheng[3]

1 Department of Advanced Design and Systems Engineering, City University of Hong Kong, Hong Kong, China

2 Faculty of Social Science, Center for Housing Innovations and Institute of Asia-Pacific Studies, Chinese University of Hong Kong, Hong Kong, China

3 China University of Geoscience, State Key Laboratory for Geological Processes and Mineral Resources, Beijing, China

*Address all correspondence to: yuanzhizhang@hotmail.com

IntechOpen

References

[1] Cruden, D.M. (1991). A simple definition of a landslide. Bulletin of the International Association of Engineering Geology, *43*, 27– 29.

[2] Varnes, D. J. (1978) Slope Movement Types and Processes. In: Schuster, R.L. and Krizek, R.J., Eds., Landslides, Analysis and Control, Transportation Research Board, Special Report No. 176, National Academy of Sciences, pp. 11-33.

[3] Alimohammadlou, Y., Najafi, A., & Yalcin, A. (2013). Landslide process and impacts: A proposed classification method. Catena, *104*, 219-232. https://doi.org/10.1016/j.catena.2012.11.013

[4] Bommer, J. J., & Rodriguez, C. E. (2002). Earthquake-induced landslides in Central America. Engineering Geology, *63*(3-4), 189-220.

[5] Keefer, D. K. (2002). Investigating landslides caused by earthquakes – A Historical review. Surveys in Geophysics, *23*, 473-510. https://doi.org/10.1023/A:1021274710840

[6] Pallister, J. S., Hoblitt, R. P., Crandell, D. R., & Mullineaux, D. R. (1992). Mount St. Helens a decade after the 1980 eruptions: Magmatic models, chemical cycles, and a revised hazards assessment. Bulletin of Volcanology, *54*, 126-146. https://doi.org/10.1007/BF00278003

[7] Donnini, M., Napolitano, E., Salvati, P., Ardizzone, F., Bucci, F., Fiorucci, F., ... Guzzetti, F. (2017). Impact of event landslides on road networks: A statistical analysis of two Italian case studies. Landslides, *14*(4), 1521-1535. https://doi.org/10.1007/s10346-017-0829-4

[8] Deng, Q., Fu, M., Ren, X., Liu, F., & Tang, H. (2017). Precedent long-term gravitational deformation of large scale landslides in the Three Gorges reservoir area, China. Engineering Geology, *221*, 170-183. https://doi.org/10.1016/j.enggeo.2017.02.017

[9] Li, Y., & Mo, P. (2019). A unified landslide classification system for loess slopes: A critical review. Geomorphology, 340, 67-83. https://doi:10.1016/j.geomorph.2019.04.020

[10] Di Traglia, F., Bartolini, S., Artesi, E., Nolesini, T., Ciampalini, A., Lagomarsino, D., ... Casagli, N. (2018). Susceptibility of intrusion-related landslides at volcanic islands: The Stromboli case study. Landslides *15*, 21-29. https://doi.org/10.1007/s10346-017-0866-z

[11] Pelinovsky, E., & Poplavsky, A. (1996). Simplified model of tsunami generation by submarine landslides. Physics and Chemistry of the Earth, *21*(1-2), 13-17. https://doi.org/10.1016/s0079-1946(97)00003-7

[12] Guzzetti, F., Mondini, A. C., Cardinali, M., Fiorucci, F., Santangelo, M., Chang, K. T. (2012). Landslide inventory maps: New tools for an old problem. Earth-Sci Rev, *112*(1-2), 42-66.

[13] Quinn, P. E., Hutchinson, D. J., Diederichs, M. S., & Rowe, R. K. (2011). Characteristics of large landslides in sensitive clay in relation to susceptibility, hazard, and risk. Canadian Geotechnical Journal, *48*(8), 1212– 1232.

[14] Pourghasemi, H. R., Moradi, H. R., Aghda, S. F., Sezer, E. A., Jirandeh, A. G., & Pradhan, B. (2014). Assessment of fractal dimension and geometrical characteristics of the landslides identified in North of Tehran, Iran. Environmental Earth Sciences, *71*(8), 3617– 3626.

[15] Zhu, M., He, Y., & He, Q. (2019). A review of researches on deep learning in remote sensing application. International Journal of Geosciences, *10*(01), 1-11.

[16] Shahabi, H., & Hashim, M. (2015). Landslide susceptibility mapping using GIS-based statistical models and remote sensing data in tropical environment. Scientific Reports, *5*(1). https://doi.org/10.1038/srep09899

[17] Morgenstern, N. R. (1997). Toward landslide risk assessment in practice. In: Cruden, D., Fell, R. (Eds.), Landslide Risk Assessment. Balkema, Rotterdam, pp. 15-23.

[18] Morgan, G. C., Rawlings, G. E., & Sobkowicz, J. C. (1992). Evaluating total risk to communities from large debris flows. Proceedings of 1st Canadian Symposium on Geotechnique and Natural Hazards. BiTech Publishers, Vancouver, BC, Canada, pp. 225 – 236.

[19] Choi, K. Y., & Cheung, R. W. M. (2013). Landslide disaster prevention and mitigation through works in Hong Kong. Journal of Rock Mechanics and Geotechnical Engineering, *5*(5), 354-365. https://doi.org/10.1016/j.jrmge.2013.07.007

[20] Powell, G. E., Tang, K. W., & Au-Yeung, Y. S. (1990). The use of large diameter piles in landslip prevention in Hong Kong. In: Proceedings of the tenth Southeast Asian geotechnical conference; pp. 197-202.

[21] Mittal, S., Garg, K. G., & Saran, S. (2006). Analysis and design of retaining wall having reinforced cohesive frictional backfill. Geotechnical and Geological Engineering, *24*(3), 499-522. https://doi.org/10.1007/s10706-004-5153-9

[22] Kundu, J., Sarkar, K., & Singh, T. N. (2017). Static and dynamic analysis of rock slope – A case study. Procedia Engineering, *191*, 744-749. https://doi.org/10.1016/j.proeng.2017.05.240

Section 2

Landslide Mechanisms and Dynamics

Chapter 2

The Effect of Aspect on Landslide and Its Relationship with Other Parameters

Seda Cellek

Abstract

Aspect is one of the parameters used in the preparation of landslide susceptibility maps. The procedure of this easily accessible and conclusive parameter is still a matter of debate in the literature. Each landslide area has its own morphological structure, so it is not possible to make a generalization for the aspect. In other words, there is no aspect in which landslides develop in particular. Generally, landslides occur in areas facing more than one direction. The biggest reason for this is that those areas are under the influence of other parameters. Therefore, it is wrong to evaluate the aspect, alone. Since it is a part of the system, it should be evaluated together with other conditioning factors. In this research, many landslides susceptibility studies have been investigated. The directions and causes of landslides have been determined from the studies. In addition, the criteria of the used aspect classes have been investigated. In the literature, the number of class intervals chosen, and their reasons were investigated, and the effects of this parameter were tried to be revealed in new sensitivity studies.

Keywords: Landslide, susceptibility, aspect, parameter, classification

1. Introduction

There are many different definitions of aspect in the literature. These definitions are made in three ways: by direction, by maximum variation, and by degree. The first definition group is the most commonly used. The concept of direction is to the come to the forefront. According to some researchers, the aspect at a point on the land surface is the direction that the tangent plane passing through that point faces and is expressed in degrees (the angle defined in the clockwise direction from the north) [1]. In its simplest form, the aspect is a data type that expresses the geographical direction in which the slopes develop.

According to the second definition, the aspect represents the maximum slope direction of the land surface [2]. Or, for any point, the aspect represents the direction of the maximum variation of the degree of variation of the height value [3]. According to some researchers, it is defined as the compass direction of the maximum rate of change [4, 5]. According to some researchers, it can also be defined as the slope direction, which defines the downward direction of the maximum rate of change in maximum, or as the dip direction, which defines the downward slope direction of the maximum altitude change rate [6, 7].

According to the third and last definition, the expression of the directions in degrees is in the foreground. Aspect defined it as the clockwise faces of a slope varying between 0^0 and 360^0, measured in degrees from the north [8, 9]. Generally, the aspect ranges from 0° to 360° and are handled as 45° groups, and the directions are grouped clockwise as north, northeast, east, southeast, south, southwest, west and northwest.

An aspect map shows both the direction and grade of a terrain at the same time. Therefore, it is an important factor in the analysis and production of landslide susceptibility maps. In the literature, there are many studies that accept and use aspect, landslide, as the main conditioning factor [2–12]. While some authors [13–16] consider landslides as a controlling factor, others [17, 18] do not see it as a conditioning factor. While some researchers say that aspect has no significant effect on landslides [19], some researchers have also argued that there is an important relationship between slope aspect and landslide occurrence [20]. According to most researchers, aspect has an indirect effect on landslide [21]. While some researchers associate this relationship mainly with precipitation [22–31], others have associated this with the general morphological trend of the area [27, 32]. According to most researchers, it has been argued that the relationship between landslide and aspect is also related to the dominant wind direction [33–35]. Some researchers, on the other hand, consider the effect of the aspect on the landslide, the general precipitation direction of the region, freeze–thaw, sunlight [35], longer snow retention on sun-drenched slopes, moisture retention, soil type, permeability, porosity, moisture, organic components, land and vegetation (forest, grassland, bushland, farmland), evapotranspiration [36], evaporation transpiration, climatic season, rock structure [37], It explains that factors such as discontinuities and fault orientation decrease the slope stability [10, 11, 24, 28, 30, 32, 38, 39]. Many parameters are used in landslide susceptibility studies, but it is stated that there are very few parameters that are thought to have a direct effect on landslides. The aspect parameter has also been investigated for a long time [3, 16, 28, 40–43], but it is one of the parameters on which no consensus can be reached [3, 44–47]. In the examined studies, it was determined that the aspect parameter indirectly affects the landslide. It is thought that this parameter triggers the landslide together with other parameters. Some researchers, especially in their studies on small-scale landslides, have determined that the angle with the slope affects the stability negatively [48–50]. Many researchers state that aspect is as effective as slope in the formation of landslides [11–13, 23, 24, 28, 30, 45, 51–55]. Apart from slope, aspect is one of the most important parameters in preparing hazard and zoning maps [13, 23, 24, 28, 30, 54].

As seen from the studies examined, the aspect parameter is a parameter that differs in each study area. For this reason, it has been interpreted that it should be examined together with other parameters rather than being an effective parameter in terms of landslide susceptibility alone [46]. According to Ramakrishnan et al. [56] stated in their study that different types of mass movements (plane, wedge, slope and soil slide) play an important role in control. However, there is no determination as to the extent to which the bee affects the landslide susceptibility.

In studies, landslides must be concentrated on slopes with a certain orientation in order to take into account the aspect. In many studies, researchers have determined that landslides are concentrated on slopes with certain orientations in their statistical evaluations [13, 22–30, 51, 57]. However, there are studies using the parameter in studies conducted in areas with equal landslide distribution in all directions. Generally, in such a finding, the lowest score is given to the aspect parameter.

The aspect factor is controlled by the climate process. Elevation and slope angle are also effective factors on this parameter. On the other hand, there are processes controlled by the aspect factor. The most important of these is plant ecology. This is followed by forestry, site selection and planning. Land morphology is under the influence of structural elements. It takes a long time to change. The biggest factor controlling the view is the structural and dynamic morphological conditions that form the silhouette of the field from past to present [58].

Although there are parameters that are agreed upon among researchers in the literature, look is not among them. For this reason, with this study, the relationship between aspect and landslide was tried to be revealed and this uncertainty in the literature was examined.

2. Effect on other parameters and landslide

It is stated that the parameter contributes to the landslides by affecting other parameters. Since wind direction causes precipitation intensity and erosion of sun-facing slopes, aspect indirectly affects landslide [33, 34]. Although it is stated in the literature that the effect increases with the angle of slope and elevation, the effect on landslides is mostly mentioned together with the climatic conditions. Aspect parameter is generally in close relationship with climatic conditions [59]. The parameter determines the effect of rain direction, amount of sunlight, solar heat, soil moisture, wind and air dryness [39, 60]. Since it controls the soil moisture concentration with the effect of climate, it is considered as an important factor indirectly triggering landslides [61–63]. Therefore, due to its morphology, how the aspect factor affects the climatic parameters by modifying it should be correlated.

The conditions for the slopes facing different directions to be affected by atmospheric events such as precipitation, sun, light, freeze–thaw are also different. Therefore, it is possible to evaluate the relationship of the parameter with the climate in 3 parts. These are precipitation, sun and wind.

2.1 Aspect-precipitation relationship

The most important factor affecting aspect is precipitation. Most of the researchers studying the aspect parameter associated landslide with precipitation. In the literature, there are studies that argue that slopes that receive precipitation and are in the shade are more susceptible to landslides. In the literature, there are researchers who stated that landslides are very common on the slopes where monsoon precipitation falls more frequently in the study areas [2, 35, 64–66]. After exposure to physical weathering during the dry season, they are prone to landslides with the emergence of strong monsoon precipitation and winds [67]. In their study in Greece, Alexakis et al. [68] and Kouli et al. [69] determined that the slopes facing northeast and northwest received heavy rainfall and the most landslides were observed here.

If precipitation exceeds the threshold value in an area and the area is unstable, landslides are likely to occur. In this respect, precipitation should be considered as a triggering factor and aspect as a preparatory factor. Critical slope angle values of soils in dry and saturated conditions are examined. It has been determined that the saturation or dryness of the soil affects the critical slope angle by about 40%. In this case, the slopes receiving the most precipitation were considered the most dangerous, and the slopes receiving the least precipitation were considered the least dangerous [27, 70].

The reason for the fact that landslides are significantly higher on a slope facing any direction compared to the others is that the torrential rains and heavy rains that developed during the landslide occurred along a line from that direction.

For this reason, it can be observed that landslides are more intense on slopes that receive heavy rainfall. This depends on the infiltration capacity, which is controlled by many factors such as the type of soil, its permeability, porosity, moisture and organic matter content, vegetation and the season in which precipitation occurs. Slopes that receive precipitation reach saturation more quickly and cause higher pore water pressure to develop within the soil. As a result, the pore water pressure on these slopes increases [11, 42, 67, 71, 72].

2.2 Aspect-snow water/freeze: thaw relationship

It has been determined that there is a negative effect on the landslide mechanism in the form of the reason that the snow cover stays longer in the places that are not exposed to the sun and the water holding capacity increases accordingly [20, 73–76]. Avcı [76] determined that in the Esence Stream Basin, which is the study area, the south-facing slopes receive plenty of precipitation with the effect of the facade systems, this precipitation falls in the form of snow in the winter season, and the increase in the amount of snow melts and precipitation in the spring season facilitates the landslides.

Landslides occurring in a certain slope direction are associated with long-term freezing and thawing movements [20, 73, 77]. In certain directions it is associated with increased snow concentrations and thus longer times for freeze and thaw action and intense erosion [77].

2.3 Aspect- solar radiation and wind relationship

Calligaris et al. [78] defined the aspect as the reflection of the sun's insolation. Aspect affects solar radiation and therefore temperature. Aspect affects the amount of heat energy taken from the sun and thus water loss by transpiration and evaporation [79]. The slopes that are most exposed to the sun's rays reveal evapotranferance [9]. This affects the soil moisture in the ground. In addition, evaporation affects vegetation distribution and type. In the literature, there are researchers who determined that landslides occur more intensely on slopes that are more exposed to sunlight [9, 11, 35, 39, 42, 71, 72, 80, 81]. In the literature, there are studies that determine that slopes that receive sun are more prone to landslides than slopes that receive rain. Bijukchhen et al. [82] determined that in their study areas, in general, slopes sloping towards the sunlight and precipitation region have a higher landslide hazard propensity compared to the slope in the rain shadow. Although this parameter is usually evaluated together with the aspect, Görüm [83] determined in her literature research that 72 studies used aspect and 3 studies used sun exposure as an input parameter.

Remondo et al. [84], on the other hand, used the values on this date in their studies for landslide susceptibility assessment, since 21 March will be the most sun exposure. Tasoglu et al. [85], in their work; they determined that it was exposed to direct sunlight in east, southeast, south and southwest directions and sunlight was quite effective in inducing landslides.

Like exposure to sunlight, the drying wind also controls soil moisture concentration. This is a determinant of landslide occurrence [61, 62, 67, 71]. Slope exposure shows possible effects of prevailing winds, differential weather and related effects.

2.4 Aspect-geology relationship

Lithology: indirectly, it triggers the landslide together with the view. Afungang et al. [86] determined that thick pyrolastics as debris in the study areas were more susceptible to landslides in windward slope directions. Yeşiloğlu [87] evaluated the effects of lithology and landslide together in his study. An aspect map has been created to be used in the evaluation of the relationship between the production of debris material from limestones and aspect. According to Ayalew et al. [70] stated in their study that the distribution of landslides in regions close to the oceans increases with the effect of wave effect, weathering and subsequent coastal erosion.

Along with the fault, there are also those who research the effect of the landslide on the landslide, there are also those who research the effect of the landslide on the landslide. There are researchers who observed that landslides intensified in certain slope directions before and after the earthquake in the study areas [2, 39, 88, 89].

Guillard and Zezere [90] stated that south-facing slopes receive more sunlight than north-facing slopes in their study area, but since the geological structure of the area is characterized by a monocline dipping to the south and southeast, more landslides occur on south-facing slopes.

2.5 Aspect- vegetation cover relationship

Aspect plays an important role in stability assessment; because it controls vegetation distribution, type, density and root growth on a land [11, 39, 80, 91]. It also controls moisture content in soil and vegetation growth due to exposure to sunlight, which also affects soil strength, landslide, infiltration and run-off rates [63, 92]. Dahal [93] added aspect data in his research for the purpose of detecting plant propagation and increasing the accuracy rate according to the aspect effect in the study area.

Champati ray et al. [94] and Srivastava et al. [95] found that most of the south-facing slopes in the Himalayan study areas were devoid of or have insufficient vegetation due to low soil moisture, which plays an important role in the assessment of slope stability in their field. On the other hand, the north side is less exposed to the sun's rays, thus conserving the moisture in the soil. For this reason, taller trees are growing, which tends to stabilize the northern slope. The absence of vegetation provides the slope material with dryness and therefore reduces its adhesion strength.

3. Relation of aspects to each other

During the literature review, it was determined that while more intense landslides were observed on the slopes facing one direction, less landslides were observed on the opposite side of this direction. Since the "south, southeast, southwest and west" aspects are generally warmer in Turkey, they are called sunny aspects. On the contrary, "north, northeast, northwest and east" aspects are also called shaded aspects because they are cooler. The sun exposure times of these two groups differ markedly. Since the slopes facing south and west are more exposed to sunlight, evaporation is rapid in these regions. Otherwise, since evaporation is slow and the soil stays moist for a long time, the risk of flooding is higher on north and east facing slopes in case of excessive precipitation [96]. Again, in his field study in Turkey, Ozsahin [97] determined the probability of the highest landslide occurrence as N and W directions and stated that the humidity was relatively higher on the slopes facing these directions.

3.1 South (S)

In areas where landslides occur on the south side, a higher amount of solar insulation occurs. On slopes with higher insulation and higher temperatures, erosion increases. Areas where vegetation is removed are exposed to direct sunlight, creating drier soil conditions, which increases the likelihood of landslides [98]. According to Devkota et al. [47], Hong et al. [99] and Chena et al. [11], most of the landslides occurred on the slopes facing south and southeast in the study areas. The biggest reason for this is that the highest precipitation rate is seen on the south-facing slopes. Meinhardt et al. [65] determined that the water saturation of the slopes increased with the effect of southwest monsoon rains in the study areas and the highest slip density was found in the south and southwest. Tombus [100], on the other hand, determined in his study that the erosion value is higher on slopes facing south than on slopes facing other directions.

3.2 North (N)

In the studies conducted in the Black Sea, it was observed that landslides were intense on the slopes facing north. The reason is that the region is under the influence of precipitation from the north and north-facing slopes are more affected by precipitation. From this, it can be concluded that the air currents coming from the sea in the study areas close to the sea will affect more areas in the region. It is known that the Black Sea receives more precipitation than the north due to the high evaporation of precipitation. For this reason, north-facing slopes are examined as the most dangerous in terms of soil saturation in the study area, and south-facing slopes are examined as the least dangerous. [101, 102]. According to Hadji et al. [9] determined that the slopes in the study area are mostly in the north-facing directions. In addition, they determined that the most precipitation in winter comes from the northwest. They also determined that they affect the clays in the ground and therefore trigger landslides.

3.3 South (S)-North (N)

In their study, Lineback et al. [103] found more landslides in the north and northwest-facing directions than in the south-facing directions. They stated that the southern parts remained drier as the reason for this. Wang and Unwin [104], on the other hand, found evidence in their study that the probability of slipping increases in the north-facing slope direction. As justification, they showed that the main precipitation directions in the Zagros Mountain Belt are north and west, and the main solar direction is east and south [105]. According to Saha et al. [4] determined that, in general, south-sloping slopes have less vegetation density than north-facing slopes, and therefore they are more sensitive to landslide activity in the study areas. On the other hand, Marston et al. [106] observed that, due to geographical conditions, north and west facing slopes have a higher moisture content for a longer period of time and cause higher landslide susceptibility in their study area. They emphasize that exposed soil on south-facing slopes is subject to cycles of wetting and drying, thereby increasing landslide activity in the Himalayas [20]. According to Rahman et al. [79] found that south-facing slopes were more exposed to the sun and north-facing slopes were least exposed to the sun in their study area.

As a result, they determined that the north direction and the least south direction were sensitive to landslides in their fields. They showed that the reason for this is that it takes longer time for the soil to dry in the shaded areas on rainy days. According to Akinci et al. [107] found that in the study areas, the slopes are more north-oriented and again, landslides occur mostly in this direction. They stated that these slopes are more humid with the effect of aspect, while the temperature and evaporation are low on the slopes facing north, and the soil moisture is high.

In addition, they stated that the amount of precipitation and snow melts are high on the southern slopes. Afungang et al. [86] found that north and northwest-facing slopes at higher altitudes received more precipitation and sun than south-facing slopes. Therefore, it was determined that the southwest-facing slopes were drier, less windy, and received less solar radiation with less landslides. Champati ray et al. [94] and Srivastava et al. [95], in their study in Himelaya, found that more landslides occurred on the southern front compared to the northern front. Temiz [101] and Yalçın [102], on the other hand, determined that north-facing slopes were the most dangerous in terms of soil saturation in the study area, and south-facing slopes were the least dangerous.

4. Aspect classes

The reasons for the change in the number of class intervals can be counted as the slopes being oriented in a certain direction, the absence of landslides in some directions or the presence of very few pixels. It is usually given to flat areas such as lakes and seas [20]. For example, the probability of landslides in "flat" areas is almost zero [34]. However, Yeşilnacar and Topal [108] with Çevik and Topal [109] stated that the landslides in the study area occurred equally in different slope orientations and emphasized that it is not an effective parameter in their studies. Aspect is measured clockwise towards north and takes positive values between 0 and 360 degrees. Aspect is measured clockwise towards north and takes positive values between 0 and 360 degrees. In order to create a slope orientation map, on the basis of 4 main geographical directions and these main directions (NE, NW, SE and SW), which of these directions the slopes face in the study area and their relations with the directions of the landslides are determined [101, 102]. It indicates 0° north, 90° east, 180° south and 270° west [32]. In the landslide analysis, a categorical structure is formed according to 45^0 angles. When the researchers grouped the slope orientation values in their studies, they determined which orientations the landslides intensified. The perspective angles and values made in the studies are given in **Table 1**.

In studies, very different grade ranges from 4 to 10 are used. According to the literature, the most preferred 8 grade ranges.

Some researchers preferred to use 4 main aspects in the aspect parameter they used in their studies. There are researchers who use the aspects divided into 4 groups in their studies in different ways. According to Temesgen et al. [110] used 4 cardinal directions: north, south, east and west. Özşahin and Kaymaz [111] have 4 classes; they used it by arranging it as straight/N-NE-NW/S-SE-SW/E-W. There are studies that use the aspect by classifying it in 5 ways [6, 97, 105, 112].

In the literature, three different directions were found in the 5-category. The first of these; flat (−1°), north (315°-360°, 0°-45°), east (45°-135°), south (135°-225°) and west (225°-315°) [113]. The second classification is; (1) SW 181^0–225^0, (2) SE 136^0–180^0, (3) ESE 91^0–135^0 and SWW 226^0–270^0, (4) NEE 46^0–90^0 and WNW

North	Northeast	East	Southeast
0^0–22.5^0, 337.5^0–360^0	22.5^0–67.5^0	67.5^0–112.5^0	112.5^0–157.5^0
South	Southwest	West	Northeast
157.5^0–202.5^0	202.5^0–247.5^0	247.5^0–292.5^0	292.5^0–337.5^0

Table 1.
Slope directions and angles.

271^0–315^0, (5) NNE 0^0–45^0 and NWN 316^0–360^0 [74]. The third and final classification is; It is flat, NE, SE, SW and NW [50].

Aspect maps divided into 6 classes are very common in the literature. Kumtepe et al. [114] prepared this classification as 0–60°, 60–120°, 120–180°, 180–240°, 240–300°, 300–360°.

The second most preferred classification in the literature is 8 classes prepared with groups of 45^0 divided into equal class intervals [35, 43, 45, 54, 65, 79, 94, 95]. This classification; N (337.5–22.5), NE (22.5–67.5), E (67.5–112.5), SE (112.5–157.5), S (157.5–202.5), SW (202.5–247.5), W (247.5–292.5) and NW (292.5–337.5) [37]. Ramakrishnan, et al. [56], on the other hand, arranged the 8-class classification differently as 45–90, 90–135, 135–180, 180–225, 225–270, 270–155 and 315–360 degrees.

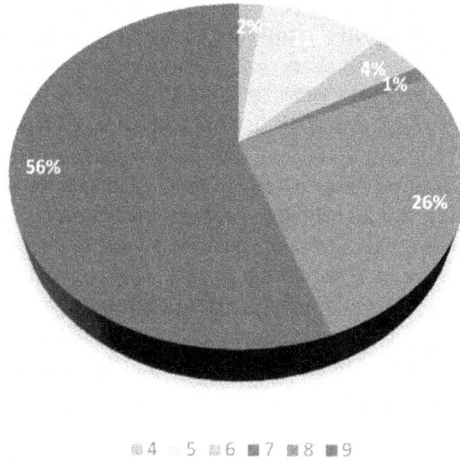

Figure 1.
Distribution of class range values used according to the literature.

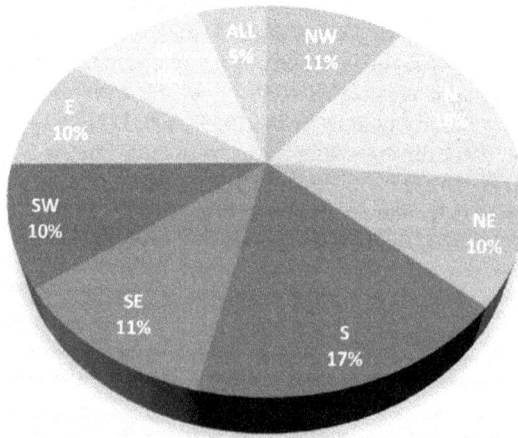

Figure 2.
Distribution of landslide areas according to directions.

According to the literature, the most preferred classification is groups of 9 [11, 32, 47, 52, 53, 66, 68, 69, 71, 87, 93, 95, 109, 102, 113]. In studies, this classification is; flat area (−1°), north (337.5° -22.5°), northeast (22.5° -67.5°), east (67.5° -12.5°), southeast (112.5° -12.5°) 157.5°), south (157.5° -202.5°), southwest (202.5° -247.5°), west (247.5° -292.5°), and northwest (292.5° -337.5°) [49, 46, 67, 73]. According to Rozos et al. [74] is this group; They used NNE, NEE, SEE, SSE, SSW, SWW, NWW, NNW, as flat shapes. The interesting thing about this classification is that the surface is displayed from 2 different angles.

The graph in **Figure 1** was prepared using the literature data. It is seen that the most used classification is the groups of 56% and 9 percent. Again, it is seen from the graph that the group of 1 to 4 is the least used class.

According to the literature, the most used direction classes are given in **Figure 2**. The direction of the landslide areas varies according to the study areas. However, in the studies examined, it is understood that the directions where landslides occur most are the slopes facing south and west. The probability of landslides in other directions is almost equal. In some studies, landslides were encountered at an equal level in all directions.

5. Conclusions

In this study, the use of aspect parameter in landslide susceptibility studies and its effect on landslide were investigated. It is one of the parameters that cannot be agreed upon by the researchers. While some researchers associate landslide occurrences in the study area with this parameter, some researchers argued that landslides are equally distributed in all directions and that the parameter is ineffective.

It is a fact that this parameter should not be evaluated alone, as in other parameters. The parameter is the predisposing factor for the triggers. One of these triggers is precipitation. There are many studies showings that intense landslides occur on slopes that receive rainfall. Climatic events such as sun, wind, snow water, freeze–thaw are also associated with the aspect parameter. The other two parameters most associated with climatic factors are geology and vegetation.

The other subject discussed in the study is the relationship of the directions with each other and with the landslide. The most common landslides seen in the studies examined are south and north directions. There is an opposite relationship between them. If there are frequent landslides on the south-facing slopes, there are almost no landslides on the north-facing slopes. Again, on the contrary, if landslides are concentrated on the north-facing slopes, landslides are not expected in the southern part. If a landslide occurs more in the south, it is associated with sun exposure, drought and lack of vegetation. Those occurring in the north are mostly evaluated by heavy rainfall, humidity and the water holding capacity of the soil.

Finally, the class ranges used in the literature are included in the study. Aspects used in the literature. In the studies, this classification is; flat area (−1°), north (337.5° -22.5°), northeast (22.5° -67.5°), east (67.5° -12.5°), southeast (112.5° -12.5°) 157.5°), south (157.5°) ° -202.5°), southwest (202.5° -247.5°), west (247.5° -292.5°) and northwest (292.5° -337.5°). Depending on the user's preference, some prefer the main classes, while others include intermediate aspects in their work. Some studies do not include aspects that do not appear to have landslides in their studies. In this way, various classifications such as 4, 5, 6, 8 and 9 are used. While the most preferred 9 classes are the least preferred groups of 4. With this study, the use of the aspect parameter in landslide susceptibility studies and its effect on the landslide together with other parameters were revealed.

Author details

Seda Cellek
Kirsehir Ahi Evran University, Kirsehir, Turkey

*Address all correspondence to: sedacellek@ahievran.edu.tr

IntechOpen

References

[1] Yomralioglu, T., Handbook of Geographic Information Systems: Basic Concepts and Applications, 2009. 480 p, ISBN 975-97369-0-X, Istanbul.

[2] Tanoli JI, Ningsheng C, Regmi AD, Jun L. Spatial distribution analysis and susceptibility mapping of landslides triggered before and after Mw7.8 Gorkha earthquake along Upper Bhote Koshi, Nepal. Arabian Journal of Geosciences. 2017; 10-13. DOI:10.1007/s12517-017-3026-9.

[3] Chen SC, Chang CC, Chan HC, Huang LM, Lin LL. Modeling typhoon event-induced landslides using GIS-based logistic regression: A case study of Alishan Forestry Railway, Taiwan. Math. Prob., Eng. 2013. Available from: https://www.hindawi.com/journals/mpe/2013/728304/.

[4] Saha AK, Gupta RP, Sarkar I, Arora M K, Csaplovics E. An approach for GIS-based statistical landslide susceptibility zonation—with a case study in the Himalayas. Landslides. 2005; 2:61-69.

[5] Lee S. Application of likelihood ratio and logistic regression models to landslide susceptibility mapping using gis environmental management. Springer Science-Business. 2004;34,2: 223-232.

[6] Bourenane H, Bouhadad Y, Guettouche MS, Braham M. GIS-based landslide susceptibility zonation using bivariate statistical and expert approaches in the city of Constantine (Northeast Algeria). Bulletin of Engineering Geology and the Environment. 2015; 74:337-355.

[7] Zhuang J, Peng C, Wang G, Chen X, Iqbal J, Guo X. Rainfall thresholds for the occurrence of debris flows in the Jiangjia Gully, Yunnan Province, China. Eng. Geol. 2015;195. doi.org/10.1016/j.enggeo.2015.06.006.

[8] Lee S. Application of logistic regression model and its validation for landslide susceptibility mapping using gis and remote sensing data. International Journal of Remote Sensing. 2005; 26, 7-10: 1477-1491.

[9] Hadji R, Chouabi A, Gadri L, Rais K, Hamed Y, Boumazbeur A. Application of linear indexing model and GIS techniques for the slope movement susceptibility modeling in Bousselam upstream basin Northeast Algeria. Arabian J. Geosci. 2016; 9:3,192. doi.org/10.1007/s12517-015-2169-9.

[10] Carrara A. Multivariate methods for landslide hazard evaluation. Math. Geol. 1983; 15:3, 403- 426.

[11] Chen W, Pourghasemi HR, Kornejady A, Zhang N. Landslide spatial modeling: Introducing new ensembles of ANN, MaxEnt, and SVM machine learning techniques. Geoderma. 2017; **305**: 314-327. doi.org/doi.org/10.1016/j.geoderma.2017.06.020.

[12] Chen W, Pourghasemi HR, Kornejady A, Xie X. GIS-based landslide susceptibility evaluation using certainty factor and index of entropy ensembled with alternating decision tree models. In book: Natural Hazards GIS-Based Spatial Modeling Using Data Mining Techniques. Adv. Nat. Technol. Hazards Res. 2018; **48**.

[13] Nagarajan R, Roy A, Vinod Kumar R, Mukherjee A, Khire MV. Landslide hazard susceptibility mapping based on terrain and climatic factors for tropical monsoon regions. Bull. Eng. Geol. Environ. 2000; **58**:275-287. doi.org/10.1007/s100649900032.

[14] Fernández T, Irigaray C, El Hamdouni R, Chacón J. Methodology for landslide susceptibility mapping by means of a GIS application to the

Contraviesa Area (Granada, Spain). Natural Hazards. 2003; 30: 297-308.

[15] Santacana N, Baeza B, Corominas J, Paz A, Marturia J. A GIS–Based Multivariate statistical analysis for shallow landslide susceptibility mapping in La Pobla De Lillet Area (Eastern Pyrenees, Spain). Natural Hazards. 2003; 30: 281-295.

[16] Ayalew L, Yamagishi H, Ugawa N. Landslide susceptibility mapping Using GIS based weighted linear combination, the case in Tsugawa Area of Agano River, Niigata Prefecture, Japan. Springer-Verlag, Landslides. 2004; 1:73-81.

[17] Neuhäuser B, Terhorst B. Landslide susceptibility assessment using "weights of-evidence applied to a study area at the Jurassic Escarpment (SW-Germany). Geomorphology. 2007; 86: 12-24.

[18] Blesius L, Weirich F. Shallow landslide susceptibility mapping using stereo air photos and thematic maps. Cartography and Geographic Information Science. 2010; 37: 2.

[19] Greenbaum D, Ton Tu M, Bowker MR, Browne TJ, Buleka J, Greally KB, Kuna G, Mcdonald AJW, Marsh SH, Northmore KJ, O'connor EA, Tragheim DG. Rapid methods of landslide hazard mapping: Papua New Guinea case study. British Geological Survey: Technical Report: Wc/95/27 Overseas Geology Series, I. Eyworth, Nottingham, British Geological Survey, 1995, Available from: Https://Core. Ac.Uk/Download/Pdf/57306.Pdf

[20] Go´Meza H, Kavazoğlu T. Assessment of shallow landslide susceptibility using artificial neural networks in Jabonosa River Basin, Venezuela. Engineering Geology; 2004:78,11-27.

[21] Jimenez-Peralvarez JD, Irigaray C, El Hamdouni R, Chacon J. Building

models for automatic landslide-susceptibility analysis, mapping and validation in ArcGIS. Nat Hazards. 2009; 571 – 590.

[22] Anbalagan R. Landslide hazard evaluation and zonation mapping in Mountainous Terrain. Engineering Geology. 1992; 32: 269-277.

[23] Van Westen CJ, Bonilla JBA. Mountain hazard analysis using a PC-Based GIS. Proceedings of the 6th International Congress of Engineering Geology. 1990; 265-271.

[24] Carrara A. Cardinali M, Detti R, Guzzetti F, Pasqui V, Reichenbach P. GIS Techniques and Statistical Models in Evaluating Landslide Hazard, Earth Surface Processes and Landforms. 1991;16: 5, 427-445.

[25] Koukis G, Ziourkas C. Slope instability phenomena in Greece: A Statistical Analysis. Bulletin of International Association of Engineering Geologists. 1991; 47-60.

[26] Juang CH, Lee DH, Sheu C. Mapping Slope Failure Potential Using Fuzzy Sets. J. Geotech. Eng. Div. ASCE. ;1992: 118, 475-493.

[27] Pachauri AK, Pant M. Landslide hazard mapping based on geological attributes. Eng. Geol. 1992;32: 81-100. doi.org/10.1016/0013-7952(92)90020-Y.

[28] Maharaj R. Landslide processes and landslide susceptibility analysis from an Upland Watershed: a case study from St. Andrew, Jamaica, West Indies. Engineering Geology. 1993; 34: 53-79.

[29] Mejia-Navarro M, Wohl EE. Geological hazard and risk evaluation using GIS: Methodology and model applied to Medellin, Columbia. Bulletin of Association of Engineering Geologists. 1994; 31, 4: 459-481.

[30] Guzzetti F, Carrara A, Cardinali M, Reichenbach P. Landslide hazard

evaluation: a review of current techniques and their application in a multi–scale study, Central Italy. Geomorphology. 1999; 31: 181-216.

[31] Luzi L, Pergalani F. Slope instability in static and dynamic conditions for urban planning: the "Oltre Po Pavese" case history (Regione Lombardia-Italy). Natural Hazards. 1999; 20:57-82.

[32] Kavzoglu T, Sahin EK, Colkesen I. Landslide susceptibility mapping using GISbased multi-criteria decision analysis, support vector machines, and logistic regression. Landslides. 2014; 11:425-439.

[33] Fan JR, Zhang XY, Su FH, Ge YG, Tarolli P, Yang ZY, Zeng C, Zeng Z. Geometrical feature analysis and disaster assessment of the Xinmo landslide based on remote sensing data. J. Mt. Sci. 2017; 14(9):1677-1688. doi. org/10.1007/s10346-017-0927-3.

[34] Liu C, Li W, Wu H. Susceptibility evaluation and mapping of China's landslides based on multisource data. Natural Hazards. 2013;1477-1495.

[35] Ahmed B. Landslide susceptibility mapping using multi-criteria evaluation techniques in Chittagong Metropolitan Area, Bangladesh. Landslides. 2014;12 (6): 1077-1095.

[36] Fernandez Merodo JA, Pastor M, Mira P, Tonni L, Herreros MI, Gonzalez E, Tamagnini R. Modelling of diffuse failure mechanisms of catastrophic landslides. Computer Methods in Applied Mechanics and Engineering. 2004; 193: 2911-2939.

[37] Balamurugan G, Ramesh V, Touthang M. Landslide susceptibility zonation mapping using frequency ratio and fuzzy gamma operator models in part of NH-39, Manipur, India. Nat Hazards. 2016.doi.org/10.1186/s40677-014-0009-y 84, 465-488.

[38] Pourghasemi H R, Rossi M. Landslide susceptibility modeling in a landslide prone area in Mazandarn Province, north of Iran: a comparison between GLM, GAM, MARS, and M-AHP methods. Theor Appl Climatol. 2017;130: 609-633.

[39] Dai FC, Lee CF, Li J, Xu ZW. Assessment of landslide susceptibility on the natural terrain of Lantau Island, Hong Kong. Environmental Geology. 2001; 40:3: 381– 391.

[40] Sarkar S, Kanungo DP. GIS application in landslide susceptibility mapping of Indian Himalayas, GIS. Landslide. 2017; 211-219.

[41] Constantin M, Bednarik M, Jurchescu MC, Vlaicu M. Landslide susceptibility assessment using the bivariate statistical analysis and the index of entropy in the Sibiciu Basin (Romania). Environ. Earth Sci. 2011; 63 (2) 397-406. doi.org/10.1007/ s12665-010-0724.

[42] Pourghasemi HR, Mohammady M, Pradhan B. Landslide susceptibility mapping using index of entropy and conditional probability models in GIS: Safarood Basin, Iran. Catena. 2012; 97: 71-84.

[43] Pawluszek K, Borkowski A. Impact of DEM-derived factors and analytical hierarchy process on landslide susceptibility mapping in the region of Rożnów Lake, Poland. Nat. Hazards. 2017; 86 (2) 919-952. doi.org/10.1007/ s11069-016-2725-y.

[44] Ruff M, Czurda K. Landslide susceptibility analysis with a heuristic approach in the Eastern Alps (Vorarlberg, Austria). Geomorphology. 2008; 94: 3: 314-324.

[45] Youssef AM, Al-Kathery M, Pradhan B. Landslide susceptibility mapping at Al-Hasher Area, Jizan (Saudi Arabia) using GIS-based

frequency ratio and index of entropy models. Geosci. J. 2015;19(1)113-134.

[46] Hasekioğulları GD. Assessment of parameter effects in producing landslide susceptibility maps. Master Thesis (in Turkish) Hacettepe University, Turkey, 2011.

[47] Devkota KC, Regmi AD, Pourghase H, Yoshida K, Pradhan B, Ryu IC. Landslide susceptibility mapping using certainty factor, index of entropy and logistic regression models in GIS and their comparison at Mugling–Narayanghat road section in Nepal Himalaya. Nat. Hazards. 2013; 65:1:135-165.

[48] Lee S, Min K. Statistical analyses of landslide susceptibility at Yongin, Korea. Environmental Geology. 2001;40 (9)1095-1113. doi.org/10.1007/s002540100310.

[49] Dai FC, Lee CF. Landslides on natural terrain physical characteristics and susceptibility mapping in Hong Kong. Mountain Research and Development. 2002; 22: 1: 40-47.

[50] Timilsina M, Bhandary NP, Dahal RK, Yatabe R. Distribution probability of large-scale landslides in central Nepal. 2014; 226: 1: 236-248. Available from: https://doi.org/10.1016/j.geomorph.2014.05.031.

[51] Saha AK, Gupta RP, Arora MK. GIS-based landslide hazard zonation in the Bhagirathi (Ganga) valley, Himalayas. Int J Remote Sens. 2002; 23: 357-369.

[52] Kayastha P, Bijukchhen SM, Dhital MR, De Smedt F. GIS based landslide susceptibility mapping using a fuzzy logic approach: A case study from Ghurmi-Dhad Khola area, Eastern Nepal. Journal of the Geological Society of India. 2013;82: 249-261.

[53] Chen CW, Sait H, Oguchi T. Rainfall intensity–duration conditions for mass movements in Taiwan, Prog. Earth Planet. Sci. 2015;2: 1-13. doi.org/10.1186/s40645-015-0049-2.

[54] Saponaro A, Pilz, Wieland, Bindi D, Moldobeko B, Parola B. Landslide susceptibility analysis in data-scarce regions: the case of Kyrgyzstan. Bulletin of Engineering Geology and the Environment. 2015;74: 1117-1136.

[55] Myronidis D, Papageorgiou C, Theophanous S. Landslide susceptibility mapping based on landslide history and analytic hierarchy process (AHP). Nat. Hazards. 2016;245-263.

[56] Ramakrishnan D, Singh TN, Verma AK, Gulati A, Tiwari KC. Soft computing and GIS for landslide susceptibility assessment in Tawaghat area, Kumaon Himalaya, India. Natural Hazards. 2013; 65:315-330.

[57] Lee S, Choi J, Mi K. Probabilistic landslide hazard mapping using GIS and remote sensing data at Boun, Korea. International Journal of Remote Sensing. 2004; 25 (11) 2037-2052. doi.org/10.1080/01431160310001618734.

[58] Chuan T, Jing Z, Jingtao L. Emergency assessment of seismic landslide susceptibility: a case study of the 2008 Wenchuan earthquake affected area. Earthquake. Engineering and Engineering Vibration. 2009; 8:28: 207-217.

[59] Bednarik M, Magulova B, Matys M, Marschalko M. Landslide susceptibility assessment of the Kralovany–Liptovsky Mikulas railway case study. Phys Chem Earth Parts. 2010; A/B/C 35(3-5): 162– 171.

[60] Kumar R, Anbalagan R. Landslide susceptibility mapping using analytical hierarchy process (AHP) in Tehri reservoir rim region Uttarakhand. Journal of the Geological Society of India. 2016;87 (3) 271-286

[61] Magliulo P, Di Lisio A, Russo F. Comparison of GIS-Based Methodologies for the Landslide Susceptibility Assessment, Geoinformatica. 2009; 13:253-265.

[62] Bui D, Pradhan B, Lofman O, Revhaug I, Dick OB. Landslide susceptibility mapping at Hoa Binh province (Vietnam) using an adaptive neuro fuzzy inference system and GIS. J. Comp. Geosci. 2011;45: 199-211. doi. org/10.1016/j.cageo.2011.10.031.

[63] Yang ZH, Lan HX, Gao X, Li, LP, Men YS, Wu YM. Urgent landslide susceptibility assessment in the 2013 Lushan earthquake-impacted area, Sichuan Province, China. Nat. Hazard. 2015; 75(3)2467-2487. doi.org/10.1007/ s11069-014-1441-8.

[64] Ruff M, Czurda K. Landslide Susceptibility Analysis with a Heuristic Approach at the Eastern Alps (Vorarlberg, Austria). Geomorphology. 2008; 94: 3-4: 314-324.

[65] Meinhardt M, Fink M, Tünschel H. Landslide susceptibility analysis in central Vietnam based on an incomplete landslide inventory: Comparison of a new method to calculate weighting factors by means of bivariate statistics. Geomorphology. 2015;234: 80-97.

[66] Zhang JQ, Liu RK, Deng W, Khanal NR, Gurung DR, Ramachandra Sri, Murthy M, Wahid S. Characteristics of landslide in Koshi River Basin, Central Himalaya. Journal of Mountain Science. 2016; 1711-1722.

[67] Wang HQ, He J, Li Y, Sun S. Application of analytic hierarchy process model for landslide susceptibility mapping in the Gangu County, Gansu Province, China. Environ. Earth Sci. 2016; 75: 422. doi. org/10.19111/bulletinofmre.502343.

[68] Alexakis DD, Agapiou A, Tzouvaras M, Themistocleous K, Neocleous K, Michaelides S, Hadjimitsis DG. Integrated use of GIS and remote sensing for monitoring landslides in transportation pavements: the case study of Paphos area in Cyprus. Natural Hazards. 2014;72: 1: 119-141,

[69] Kouli M, Loupasakis C, Soupios P, Rozos D, Vallianatos F. Landslide susceptibility mapping by comparing the WLC and WofE mutli-criteria methods in the West Crete Island, Greece. Environ Earth Sci. 2014.

[70] Ayalew L, Yamagishi H. The application of GIS-based logistic regression for landslide susceptibility mapping in the Kakuda-Yahiko Mountains. Central Japan. Geomorphology. 2005; 65: 1-2: 15-31.

[71] Pham BT, Bui DT, Prakash I, Dholakia MB. Hybrid integration of multilayer perceptron neural networks and machine learning ensembles for landslide susceptibility assessment at Himalayan area (India) using GIS. Catena 2017;149: 52-63. htdoi. org/10.1016/j.catena.2016.09.007.

[72] Kritikos T, Davies T. Assessment of rainfall-generated shallow landslide/ debris-flow susceptibility and runout using a GIS-based approach: Application to western Southern Alps of New Zealand. Landslides. 2014;12 (6)1051-1075. doi.org/10.1007/ s10346-014-0533-6.

[73] Ilia I, Tsangaratos P. Applying weight of evidence method and sensitivity analysis to produce a landslide susceptibility map. Landslides. 2016; 379-397.

[74] Rozos D, Bathrellos GD, Skilodimou HD. Landslide susceptibility mapping of the northeastern part of Achaia Prefecture using Analytical Hierarchical Process and GIS techniques. Bull. Geol. Soc. Greece, 2010.Proceedings of the 12th

International Congress, Patras may, XLIII, 3, 1637-1646.

[75] Wilson JP, Gallant JC. Digital terrain analysis, Chapter 1, In., Eds. Terrain analysis: Principles and applications. New York. 2000; 1-27.

[76] Avcı V. Landslide susceptibility analysis of Esence Stream Basin (Bingöl) by weight- of- evidence method. International Journal of Social Science. 2016; 287-310.

[77] Rozos D, Pyrgiotis L, Skias S, Tsagaratos P. An implementation of rock engineering system for ranking the instability potential of natural slopes in Greek territory: an application in Karditsa County. Landslides. 2008; 5(3):261-270.

[78] Calligaris C, Poretti G, Tariq S, Melis, MT. First steps towards a landslide inventory map of the Central Karakoram National Park. European Journal of Remote Sensing. 2017; 46:1, 272-287. https://www.tandfonline.com/doi/pdf/10.5721/EuJRS20134615.

[79] Rahman G, Atta-ur-Rahman S, Collins A E. Geospatial Analysis of landslide susceptibility and zonation in Shahpur Valley, Eastern Hindu Kush using Frequency Ratio Model. Proceedings of the Pakistan Academy of Sciences: Pakistan Academy of Sciences B. Life and Environmental Sciences. 2017;54 (3): 149-163. Available from: https://www.paspk.org/wp-content/uploads/2017/09/Geospatial-Analysis-of-Landslide.pdf

[80] Sidle R, Ochiai H. Landslides: Processes, Prediction, and Land Use., Geography. 2006Book chapter, https://www.researchgate.net/publication/292653165_Landslides_Processes_Prediction_and_Land_Use

[81] Kornejady A, Ownegh M, Bahreman. Landslide susceptibility assessment using maximum entropy model with two different data sampling methods. Catena. 2017; 152:144-162, doi: 10.1016/j.catena.2017.01.010.

[82] Bijukchhen SM, Kayastha P, Dhital, MR. A comparative evaluation of heuristic and bivariate statistical modelling for landslide susceptibility mappings in Ghurmi-Dhad Khola, East Nepal. Arabian J. Geosci., 2013; 6: 2727-2743. doi.org/10.1007/s12517-012-0569-7.

[83] Görüm T. Landslide susceptibility analysis with geographic information systems and statistical methods: Melen Gorge and near vicinty. İstanbul University, Master Thesis, Istanbul (unpublished), 2006.

[84] Remondo J, Gonzalez-Diez A, Teran JRD, Cendrero A. Landslide susceptibility models utilising spatial data analysis techniques: a case study from the lower Deba Valley, Guipúzcoa (Spain). Natural Hazards. 2003;30: 267-279.

[85] Tasoglu İK, Keskin Çıtıroglu H, Mekik Ç. GIS-based landslide susceptibility assessment: A case study in Kelemen Valley (Yenice-Karabuk, NW Turkey). Environ. Earth Sci. 2016; 75: 1295. https://doi.org/10.1007/s12665-016-6098-z.

[86] Afungang RN, Nkwemoh C, Ngoufo R. Spatial modelling of landslide susceptibility using logistic regression model in the Bamenda Escarpment Zone, NW Cameroon. International Journal of Innovative Research & Development. 2017; 6 :2:187-199.

[87] Yeşiloğlu N. Eğirdir (Isparta) yerleşim merkezi için heyelan olası tehlike değerlendirmesi ve haritalaması, yüksek lisans tezi. Hacettepe Üniversitesi, Ankara, 270p, 2006.

[88] Tang C, Zhu J, Qi X, Ding J. Landslides induced by the Wenchuan earthquake and the subsequent strong

rainfall event: A case study in the Beichuan area of China. Eng. Geol. 2011; 122: 22-33. doi.org/10.1016/j. enggeo.2011.03.013.

[89] Yang ZH, Lan HX, Gao X, Li LP, Meng YS, Wu, YM. Urgent landslide susceptibility assessment in the 2013 Lushan earthquakeimpacted area, Sichuan Province, China. Nat Hazards. 2015; 2467-2487.

[90] Guillard C, Zezere J. Landslide Susceptibility assessment and validation in the framework of municipal planning in Portugal: The Case of Loures Municipality, Environmental management. 2012; 50: 721-735. Available from: https://link.springer. com/article/10.1007/s00267-012-9921-7.

[91] Sidle RC. Influence of Forest Harvesting Activities on Debris Avalanches and Flows. 1985. Available from: http://citeseerx.ist.psu.edu/ viewdoc/download?doi=10.1.1.512.8208 &rep=rep1&type=pdf.

[92] Ramesh V, Anbazhagan S. Landslide susceptibility mapping along Kolli hills Ghat road section (India) using frequency ratio, relative effect and fuzzy logic models. Environ Earth Sci. 2015; 8009-8021.

[93] Dahal RK. Regional-scale landslide activity and landslide susceptibility zonation in the Nepal Himalaya. Environmental Earth Sciences. 2014;71: 12:5145, 5164.

[94] Champati ray PK, Dimri S, Lakhera RC, Kumar Sati S. Fuzzy-based method for landslide hazard assessment in active seismic zone of Himalaya. Landslides. 2007;4(2):101-111,

[95] Srivastava V, Srivastava HB, Lakhera RC. Fuzzy gamma based geomatic modelling for landslide hazard susceptibility in a part of Tons river valley, northwest Himalaya, India. 2010; 1:3. https://www.tandfonline.com/doi/

citedby/10.1080/19475705.2010.490103? scroll=top&needAccess=true

[96] Yılmaz G. Afete duyarlı planlama kapsamında planlama jeorisk ilişkisi ve CBS ile analizi, Bartın Kenti Örneği. Yüksek Lisans Tezi, Gazi Üniversitesi, Ankara, 2008.

[97] Özşahin E. Landslide susceptibility analysis by geographical information systems: the case of Ganos Mount (Tekirdağ) (in Turkish). Electron. J. Map Technol. 2015; 7 (1) 47-63. doi. org/10.15659/hartek.15.04.68.

[98] Rajakumar P, Sanjeevi S, Jayaseelan S, Isakkipandian G, Edwin M, Balaji P, Ehanthalingam G. Landslide susceptibility mapping in a hilly terrain using remote sensing and GIS. Journal of the Indian Society of Remote Sensing. 2007; 35: 31-42. Available from: https://link.springer. com/article/10.1007/BF02991831.

[99] Hong H, Naghibi S A, Pourghasemi HR, Pradhan B. GIS-based landslide spatial modeling in Ganzhou city, China. Arab J Geosci. 2016; 9: 2.1: 26.

[100] Tombuş FE. Uzaktan algılama ve cografi bilgi sistemleri kullanılarak erozyon risk belirlemesine yeni bir yaklaşım, Çorum ili örneği. Yüksek Lisans Tezi, Anadolu Üniversitesi, Eskişehir. 2005.

[101] Ercanoglu M, Temiz AF. Application of logistic regression and fuzzy operators to landslide susceptibility assessment in Azdavay (Kastamonu, Turkey). Environmental Earth Sciences. 2011: doi:10.1007/ s12665-011-0912-4

[102] Yalçın A. Ardeşen (Rize) yöresinin heyelan duyarlılığı açısından incelenmesi. Doktora Tezi, Karadeniz Teknik University, Trabzon. 2005.

[103] Lineback GM, Marcus WA, Aspinall R, Custer SG. Assessing

landslide potential using GIS, soil wetness modeling and topographic attributes, Payette River, Idaho. Geomorphology. 2001; 37: 149-165.

[104] Wang S.Q., Unwin D.J., 1992. Modelling landslide distribution on loess soils in China: an investigation. International Journal of Geographical Information Systems 6:391-405.

[105] Tangestani MH. Landslide susceptibility mapping using the fuzzy gamma approach in a GIS, Kakan catchment area, southwest Iran. Australian J. Earth Sci. 2004; 51: 439-450. doi.org/10.1111/j.1400-0952.2004.01068.x.

[106] Marston R, Miller M, Devkota L. Geoecology and mass movements in the Manaslu Ganesh and Langtang-Jural Himals, Nepal. Geomorphology. 1998; 26: 139– 150.

[107] Akıncı H, Kılıçoğlu C. Production of landslide susceptibility map of Atakum (Samsun) district. MÜHJEO'2015: National Engineering Geology Symposium, 3-5 September 2015, Trabzon.

[108] Yeşilnacar E, Topal T. Landslide. Susceptibility mapping: a comparison of logistic regression and neural networks methods in a medium scale study, Hendek Region (Turkey). Engineering Geology. 2005; 79: 251-266.

[109] Çevik E, Topal T. GIS-based landslide susceptibility mapping for a problematic segment of the natural gas pipeline, Hendek (Turkey). Environmental Geology. 2003; 949-962.

[110] Temesgen B, Mohammed MU, Korme T. Natural hazard assessment using GIS and remote sensing methods, with particular reference to the landslides in the Wondogenet Area, Ethiophia. Phys. Chem. Earth. 2001; 26-9: 665-675.

[111] Özşahin E, Kaymaz ÇK. Landslide susceptibility analysis of Camili (Macahel) Biosphere Reserve Area (Artvin, NE Turkey). Turkish Studies - International Periodical For The Languages, Literature and History of Turkish or Turkic. 2013; 8(3)471-493. doi.org/10.7827/TurkishStudies.4260.

[112] Caniani D, Pascale S, Sdao F, Sole A. Neural networks and landslide susceptibility: a Case study of the urban area of Potenza. Natural Hazards. 2008; 45: 55– 72.

[113] Avcı V. Analysis of landslide succeptibility of Manav Stream Basin (Bingöl). The Journal of International Social Research. 2016; 9:42-9. doi: 10.17719/jisr.20164216199.

[114] Kumtepe P, Nurlu Y, Cengiz T, Sütçü E. Bolu çevresinin heyelan duyarlılık analizi [Bildiri]. TMMOB Coğrafi Bilgi Sistemleri Kongresi, 02-06 Kasım 2009, İzmir.

Chapter 3

Implications of Soil Properties on Landslide Occurrence in Kigezi Highlands of South Western Uganda

Denis Nseka, Vincent Kakembio, Frank Mugagga,
Henry Semakula, Hosea Opedes, Hannington Wasswa
and Patience Ayesiga

Abstract

Generally, soil characteristics have a significant influence on landslide occurrence. This issue has, however, not yet been adequately analysed in Kigezi highlands of South Western Uganda. In this study, soil properties such as dispersion, grain size distribution, Atterberg limits, shear strength and clay mineralogy were analysed to establish their contribution to the spatial distribution of landslides in Kigezi highlands. The results demonstrate that deep soil profiles ranging between 2.5 and 7 meters were dominated by clay-pans at a depth between 0.75 and 3 meters. Although the uppermost surface horizons of the soil profile are loamy sand, the clay content is more than 35% especially in the sub soil. This suggests that the soil materials are Vertic in nature. In addition, the upper soil layers predominantly contain quartz, while subsurface horizons have considerable amounts of illite as the dominant clay minerals, ranging from 43–47%. The average liquid limit and plasticity index was 58.43% and 33.3% respectively. Besides, high average computed weighted plasticity index (28.4%) and expansiveness (38.6%) were obtained. These soil characteristics have great implication on the timing and nature of landslide processes in the study area. A change in soil material due to varying moisture content is thought to be a major trigger of landslides in Kigezi highlands of South Western Uganda. This understanding of soil characteristics is a key step in mitigating landslide hazards in the area.

Keywords: soil properties, landslide occurrence, Kigezi highlands

1. Introduction

Landslides are among the major life threatening global natural disasters that cause great environmental and developmental challenges [1]. Landslides normally occur on terrains with steep slopes [1, 2]. Slope failures are attributed to factors like topography, geotechnical properties of the material as well as the existence of discontinuities [3, 4]. Most of these factors are specific to particular areas and thus, site-specific studies are important [2]. In Ref. [5] noted that soil types and soil texture are primary-level factors, while elevation, land cover types, and drainage density are next in terms of landslide inducement. In [6] the low permeability of fine textured clayey soils exacerbates the vulnerability to landslides. This is due to increased saturation and pore water pressure which reduces soil shear strength [1]. In [7] it is also noted that landslide occurrence is common in regions where sandy clay loams are underlain by sandy clay soils. The stability of slope materials is also influenced by the presence of swelling clay minerals [8]. Soils with high clay content are considered to be the leading cause of landslides in most East African highlands [7]. Soil materials with clay content exceeding 30% and classified as vertic soils, have swell-shrink characteristics [9] which make them highly vulnerable to landslides [10].

In Ref. [11] it is indicated that clay mineralogy affects the shear and frictional resistance of the soils. Among the clay minerals, smectite, particularly montmorillonite, and illite decrease the soil residual strength, owing to their peculiar colloid-chemical characteristic [12] and contribute to landslide occurrence [7]. Montmorillonite a swelling clay mineral, has a negative behaviour, very strong attraction for water and can induce soil collapse due to susceptibility to volume change [13]. The dispersive, collapsible and expansive nature of clay soils constitutes what is referred as problem soils [14]. In [15] it is noted that problem soils are widespread around the world, notwithstanding the little attention they receive in many landslide studies. At various moisture contents, landslides may be induced by problem soils due to their distinct shrink-swell properties [14].

The stability of any slope is affected by specific soil parameters including bulk density, shear strength, clay mineralogy and particle size distribution [1, 7, 16]. Such soil properties vary significantly in space and require site-specific investigations to understand their contribution to landslide occurrence [1, 17]. Whereas previous landslide studies have focused on volcanic soils of Mount Elgon region in Eastern Uganda [1, 7, 18, 19], the present study examines the influence of selected morphological properties of non-volcanic soils on landslide occurrence in Kigezi highlands. The study results provide a comparative analysis of the correlation between soil properties and landslide occurrence in both mountainous as well as highland areas of Uganda.

2. Materials and methods

2.1 Study area

The study was carried out in Rukiga uplands located within Kigezi highlands of South Western Uganda. The study area is situated between 01°21′25″ and 0°58′08″ South, and 29°43′30″ and 30°05′51″ East (**Figure 1**). Rukiga uplands an area covering 427 km^2 [20], span the attitudinally heterogeneous landscapes of Rwamucucu, Maziba, Kashambya, Bubaare, Bukinda, Kamwezi and Kaharo [21] was selected for this study on the basis of its unique topography which is synonymous with reported high incidences of landslides, with visible scars unlike other parts of the highlands where landslide scars have been concealed owing to rapid regeneration [22]. The topography is very rugged, with narrow steep convex slopes and high valleys between hills. Most of these valleys have drainage lines connecting to the main valley [23]. The topography has substantial flat-topped ridges and hills, with short, steep-sided deep valleys fragmented by fluted spurs [24]. Landslide scars were, therefore, identified following an inventory with the help of local communities (**Figure 2**). The landslide scar zones provided soil sampling sites for analysis of soil characteristics and their influence on landslides (**Figure 3**).

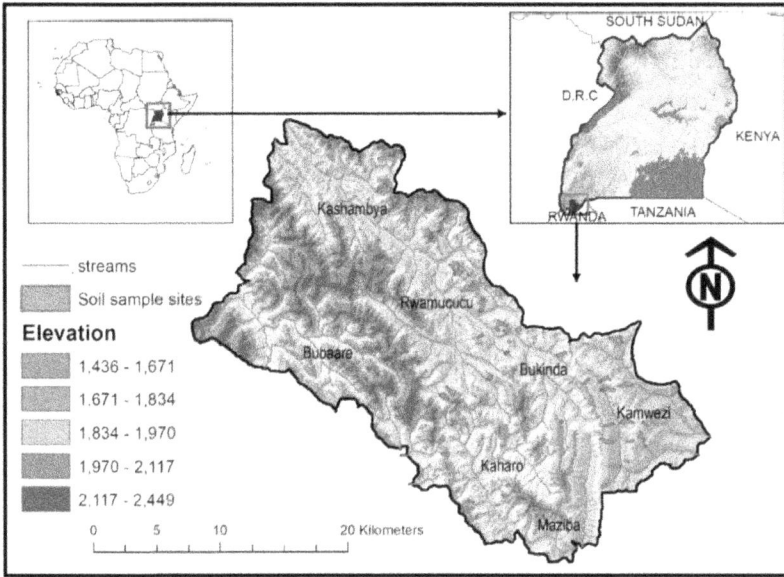

Figure 1.
Location of the study area and soil sampling sites.

Figure 2.
Field investigations to identify landslide scars and soil sampling. Photo credit, Nseka. November, 2019.

The highland's geology is sedimentary in nature of the Precambrian rock system [25], categorised by [26], as phyllite, shale, sandstone, quartzite, granite and gneisses of granitic composition. Other rock types of the study area include grades of schists like quartz-schists and fine textured mica-schists which belong to both the Ankole-Karagwe rock system as well as the Achaean basement complex [25].

Since rock types have influence on slope factors like slope angle and stability, slopes with phyllites and shales underneath are more affected by instability processes compared to those covered by quartzite and micaceous sandstones [25]. Additionally, slope sections with relatively weaker rocks beneath like shales have deep soil profiles attributed to high weathering rates [23]. Besides, slopes underlain

Figure 3.
In-situ soil property analysis within the landslide scars. Photo credit, Nseka. November, 2019.

by quartzite and granitic intrusions are less prone to landslides because they are covered by shallow soils and in some cases by bare rocks [25]. Most of the water courses within the highlands contain varied patches of alluvia sand and clay [26]. For example, coarse sands have been oberved in the warped valleys with gneiss and granite intrusions beneath, while shale and phyllite are associated with clay deposits [25]. By implication, the local geology has an important influence on landslide occurrence in the highlands. The soil types in the study area include Luvisols, Histosols, Acric Ferralsols and Dystric Regosols [23].

The climate of Kigezi highlands is warm to cool humid, characterised by a bimodal rainfall pattern with annual rainfall of 1092 mm [20] classified as moderate. Rainfall, however, increases from 1250 to 1540 mm or more in high altitude areas of greater than 2000 m above sea level [27, 28]. The main rainfall seasons span from mid-February to May with a peak in March/April, and September to December with a peak in October/November [29]. Since the highland receives significant amounts of rainfall, the study area has various highlands streams which drain valleys incised within the ridges and hills [30]. Until about a century ago, the highlands' vegetation cover was characterised by montane forests [23]. Depletion of vegetation cover has occurred in the highlands due to increased human interference [31]. The highlands of the study area are currently characterised by *Eucalyptus globulus*, *Pinus leiophylla* and farmlands [28].

2.2 Soil morphological analysis

To evaluate the soil morphology-landslide relationships, field investigations were conducted and soil samples taken at different depths and points along the slope profile and positions. In this study, 120 soil samples were collected and used in the analysis of soil morphological characteristics. Soil description was done according to FAO guidelines [32]. Visible landslide scars were identified and categorised into 10 groups based on their morphological characteristics.

Profiles in the upper slope sections were dug to a depth of 1 to 1.5 m (classified as shallow). Profiles in the middle slopes ranged between 2 to 4 m (medium) while those in the lower sections were greater than 5 m (deep). Soil samples were obtained from within and outside the landside scars. Onsite analysis of physical soil properties was conducted within landslide scars, min pits, auger holes, and full profile representative sites (**Figure 3**). Site analysis sought to characterise among other things, soil matrices and bedrock within each landslide. Soil horizons were analysed in detail with a specific focus on depth, colour, texture, presence and location of claypans, and structure.

2.2.1 Soil-water infiltration

Characterisation of soil infiltration levels is an importance aspect when analysing any landslide occurrence in a given region [26]. Quantification of infiltration rates is important to understand the mechanism of slope failure [25]. In situ infiltration tests were performed purposely to measure soil-water infiltration rates. Infiltration tests were performed at various slope positions including lower-bottom, lower-middle, upper-middle and uppermost. This was done owing to variations between topographic configurations and soil characteristics. A total of 32 experiments were performed on 4 landslide sites with different land use and cover categories. Field measurements of soil-water infiltration were done using the double ring infiltrometer method [33] consisting of two concentric metal rings [34, 35].

2.2.2 Rainfall data

The frequent heavy rainfall received within Kigezi highlands have been observed to be a main trigger of landslides because, the rains increase pore water pressure in voids [28]. Given the importance of pore water pressure and antecedent moisture in triggering landslides in the study area [23, 29], rainfall data are important in their analysis [24]. The rainfall data used in this study were obtained from Kabale Meteorology Station, weather data 2015: WMO No. 63726, National No. 91290000, station name Kabale, at Elevation 1867 m, Latitude 01°15′, Longitude 29°59″. Rainfall data were compared with landslide occurrence periods to ascertain their relationships.

2.3 Laboratory analyses

The scope of laboratory testing comprised shear box, Atterberg limits, sieve and hydrometer analysis, specific gravity using test standards and XRD analysis. For purposes of defining the dominant soil mechanical and physical properties, several laboratory tests were undertaken. The grain size distribution, unit weight, natural water content, degree of saturation and Atterberg limits were among the major physical and mechanical soil properties examined. Soil property tests were undertaken in accordance with the British Standard BS 1377 procedures [36]. These tests were used in the estimation of porosity, dispersion, saturated specific weight, and saturation humidity. These soil properties were considered relevant in understanding and characterising site-specific landslide controls ("For example, see [1, 7, 17]").

2.3.1 Determination of particle size

Particle size was determined using mechanical analysis. In this method, the size range of particles within a soil sample was determined and the results were expressed as a percentage of the total dry weight. To determine the soil particle-size distribution, *Sieve analysis* and *Hydrometer analysis* methods were used on particle sizes greater and less than 0.075 mm in diameter respectively. Sieve analysis involved shaking of the soil sample through a set of sieves (ranging from 75.00 mm to 0.075 mm). The percentage of the finest size and that of the weights retained on each of a series of standard sieves of decreasing size were used to infer particle size distribution.

2.3.2 Atterberg limits

A number of soil samples obtained from the landslide sites were used for Atterberg limit tests undertaken in the geotechnical laboratory. The tests were conducted to determine among others the clay plasticity and measure the threshold water contents of a fine soil. Atterberg limits were used to determine soil plasticity which provides a clue to the type of mass movement that would characterise a given area ("For example, see [37]"). The Atterberg limits were important in determining among others, soil strength, type, stability, behaviour, and state of consolidation. Using the procedures in ASTM Standard D 4318 and D431/84 and also CNR UNI 100141 ("For example, see [38]") the limits were determined as Liquid Limit (LL) and Plastic Limit (PL). The tests gave an indication of the levels of saturation and response of a soil material to landslides. The results of grain-size distribution and Atterberg limits tests were used to classify the colluvium soils according to the Unified Soil Classification System which enabled further classification of fine materials.

2.3.3 Soil dispersion tests

A double hydrometer test was used based on Stoke's law of settling velocity as an indicative laboratory test for identification of dispersive soils ("For example, see [11, 14]").

2.3.4 Soil expansiveness

To determine whether the soils in the study area are susceptible to slope failure, investigations were done to determine the expansiveness of the soil. Three procedures were done to determine the expansiveness of the soil in the study area. During such field surveys, a description of the crack patterns on the soil surface was done at different sites to determine the swelling-shrinking characteristics of particularly clay soils. The second procedure entailed calculating the weighted Plasticity Index (PI_w) on the fraction < 425 μm and weighted for the sample's actual content of particles < 425 μm using the formula:

$$PI_w = PI * (\%passing\ 425\mu m)/100. \tag{1}$$

Where PI is the Plasticity index.
The third procedure involved calculating the expansiveness (ε_{ex}) using the formula:

$$\varepsilon_{ex} = 2.4w_p - 3.9w_s + 32.5, \tag{2}$$

Where; w_p = (Plastic Limit) * (%passing 425 mm)/100 and w_s = (Shrinkage Limit) * (%passing 425 mm)/100.

2.3.5 Clay mineralogy analyses

Clay mineralogical composition was analysed using X-ray diffraction. Soil samples were prepared for XRD analysis using the back loading preparation method ("For example, see [39]"). A PANalytical Aeris diffractometer with PIXcel detector in combination with fixed relative phase amounts (weight %) were estimated by X'Pert Highscore plus software and Rietveld method respectively.

2.3.6 Shear strength tests

To determine material strength, shear strength tests were executed on samples obtained from the landslide sites. These tests included Shear Box and Unconsolidated – Undrained (UU) tests. The tests were conducted to determine the shear strength parameters of soil cohesion (c) and the angle of internal friction (φ) in accordance to British Standard (BS) 1377: Part 7: 1990 [36].

3. Results

3.1 Soil morphological properties

3.1.1 Soil profile characteristics

At the depth less than 0.85 m, shallow soil groups were observed and these occured on very steep (i.e. between 35° and 45°), and precipitous (i.e. >45°) slopes. Medium and deep soil groups in the range of 1.5–4 m and greater than 6 m, were found occupying midslopes along topographic hollows and lower slopes respectively. Surface layers were covered by deposited black soils with an average depth of 0.5 m to 1 m. The B and C horizons had depths ranging from 0.65 m to 3.44 m and 0.88 m and 5.7 m, with a reddish-brown colour and coarse-textured materials concentrated in the lowest layers, respectively. At a 1.3 m depth along the spur slopes, unweathered materials were observed. Other unweathered materials were observed at a depth of 4.5 m along topographic hollows and 7 m deep in the valley bottoms. Along the soil profiles, visible colours and textural gradation within 3 distinct horizons were seen. An abrupt change in colour from yellow (5Y8/2) to brown (7.5YR5/2) clay with Gs 2.9 g/cm^3 and γ 2.06 g/cm^3 was exhibited in the upper slope elements for most of the analysed profiles. Dark reddish brown (5YR 3/6) and orange (7YR 5/6) colour in moist and dry conditions respectively characterised most of the surface horizons. Surface horizons with brown (7.5YR 4/4) colour in both moist and dry conditions and a dark to dull reddish-brown colour (7.5YR 4/6) in underlying horizons dominated most of lower slope element profiles. Soil colour in combination with other physical properties, were used to differentiate horizon types of the same and different soil profiles.

A relationship was noticed between the regions topographic characteristics and the depth of soil. Basing on the soil profile description, it was noted that the depth of soil reduced with an increase in slope angle. Whereas very deep soils (>5 m) were dominant on slope sections with slope angles less than 10°, moderately deep soils (2 to 5 m) were noticed on slope angles between 15° and 35°. Slope elements with slope angles greater than 35°, were characterised with shallow thin soils of less than 1 m (Appendix 1). A relationship was also established between slope curvature and the depth of the soil. Very deep, deep and moderately deep soil profiles (>2.5 m) were dominant along concave profile elements common in topographic hollows. Convex profile elements along spur slopes and hilltops were dominated by shallow thin soils of less than 1 m (Appendix 1).

3.1.2 Location of clay pans

Field observations within dug profiles indicated the presence of claypans at a depth of 0.75 m to 3 m (**Figure 4**). Variations in the location of claypans is noticeable along the soil profile. In some profiles, the claypans were found close to the surface while in some horizons they were identified at greater depths. Whereas the upper slope soil profiles had claypans close to the surface at less than 0.85 m depth,

Figure 4.
Soil profiles dominated by claypans in the sub-surface soil horizons. Photo credit, Nseka, may 2019.

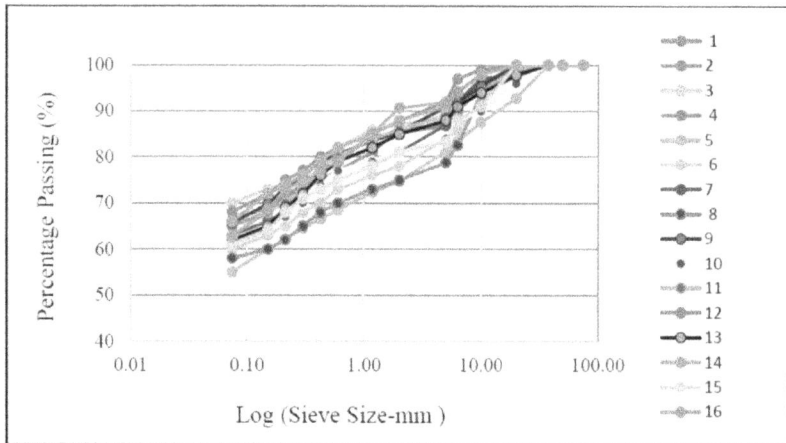

Figure 5.
Particle size distribution curves.

lower slope soil profiles had pans at greater than 2 m depth. Claypans in soil profiles along the middle slope sections existed at depth ranging between 1 and 2 m. Characterising variations in depth of claypans horizon is an important step in explaining other soil properties, as expounded in the discussion section.

3.2 Effect of particle size on landslide occurrence

The study area is dominated by clay soils in the subsurface horizons. The particle size determination (**Figure 5**) shows that soils are dominated by clay presence, except for the uppermost surface horizons.

From the materials analysed, fine grained materials of either clay or silt were predominant (**Figure 5**). The texture of the soil varied with profile depth as revealed by sieve analysis. In the subsurface horizons, finer clay and silt materials dominates, while sand particles are the most dominant in surface soils. It was observed that all surface horizons had sand (33–55%), silt (22–40% and clay (10–30%). In the deeper horizons, sand was observed to reduce drastically to less than 23%, while clay increased to greater than 50% (Appendix 1). The clay content for all the samples in the sub soil and deep soil horizons is well above the 32% threshold for vertic soils. The swelling and shrinkage characteristics of vertic soils are very important in the localisation of landslides, as explored in the subsequent section.

3.3 Clay mineralogy and landslide occurrence

The results of the X-ray diffraction tests are provided in **Figure 6** and **Table 1**.
Following the X-ray diffraction patterns, it was established that quartz, illite/muscovite and kaolinite are the dominant minerals in the soil materials. From the soil mineralogical analysis conducted, it was established that quartz was the dominant soil constituent within the top layers. By implication, these high amounts of quartz within surface soil horizons affect the behaviour of the soil to incoming rainfall as will be unravelled in the subsequent section. Considerable amounts of illite/muscovite as the dominant clay minerals, ranging between 43% and 47% (**Table 1** and **Figure 6**) were present in the subsurface soil horizons. Notwithstanding the absence of smectite clays, the soils contain large amounts of moderately expansive clays, particularly illite/muscovite.

3.4 Soil dispersion and landslide occurrence

The LL and Plasticity Index for all the tested samples are greater than 50% and 30% respectively (**Figure 7** and **Table 2**).
The fine-grained soil materials tested had LL with an average value of 58.43%, ranging between 50.43% and 66.43%, which is considered to be of high plasticity. Whereas the plasticity index ranged between 22.4% and 44.2%, the plastic value ranged between 21.3% and 28.9% (**Figure 7**). High plasticity index of more than 30

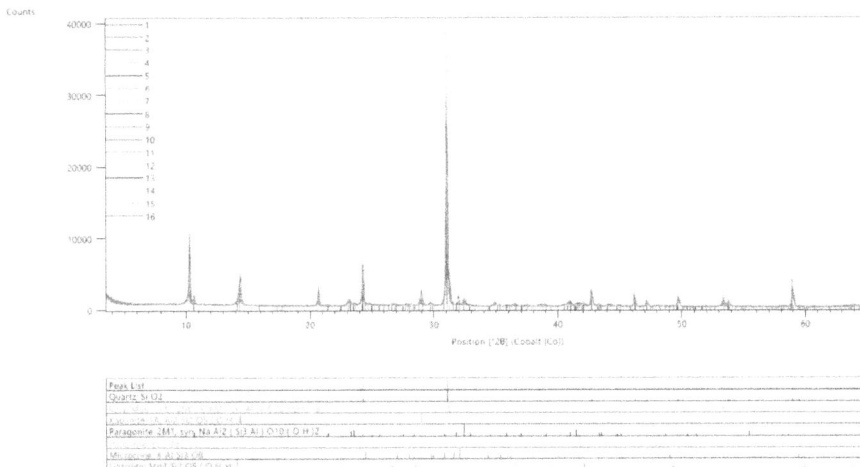

Figure 6.
XRD patterns of clay minerals.

Samples	Soil horizon	Quartz	Illite/ muscovite	Kaolinite	Paragonite	Haematite	Microcline	Lizardite
1	Top soil	56	26	9	4	1	0	4
2	sub soil	46	43	7	2	1	0	1
3	Top soil	60	27	6	6	2	0	0
4	sub soil	39	43	8	4	3	1	1
5	sub soil	42	41	10	3	2	1	1
6	sub soil	43	45	8	2	1	1	0
7	sub soil	43	42	9	3	1	1	0
8	sub soil	36	47	11	3	1	1	1
9	sub soil	36	44	12	5	1	1	1
10	Top soil	75	18	5	0	2	0	0
11	Top soil	62	28	6	2	1	0	2
12	Top soil	76	16	6	1	0	1	0
13	Top soil	71	22	5	0	1	0	2
14	sub soil	41	46	7	4	0	1	1
15	Top soil	64	22	6	4	2	0	2
16	Top soil	82	12	3	2	0	2	0

0 = n.d. – not detected above the detection limit of 0.5–3 weight percent.

Table 1.
XRD mineral distribution in percentages.

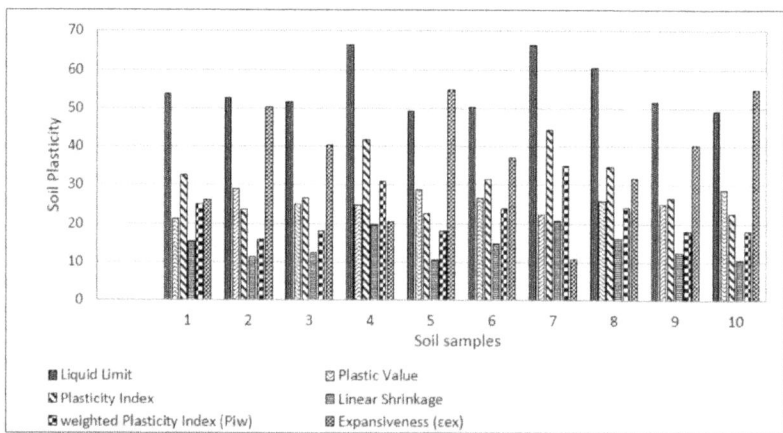

Figure 7.
Plasticity index parameters.

was detected in the clay materials, thus rendering them as highly plastic and expansive. The computed Linear Shrinkage (LS) ranged between 10.53 and 20.76. Whereas soil expansiveness (ε_{ex}) ranged between 10.7 and 54.8 averaging 38.6, the computed weighted Plasticity Index (PI_w) ranged between 17.92% and 34.92% averaging 28.4% from all the analysed soil samples (**Figure 7**). As regards to soil

Soil sample		Plasticity							Shear strength parameters		Soil class	USCS
									C′	φ′		
	LL	PL	PI	LS	w_p	w_s	PI_w	ε_{ex}	kPa	(Degree)		
1	58	26	31	15	20	11	24	37	7	3	Clayey	CH
2	54	21	32	15	16. 4	12	25	26	7	4	Clayey	CH
3	66	22	44	21	18	16	35	11	5	5	Clayey	CH
4	58	27	31	15	20	11	23	39	7	4	Clayey	CH
5	53	29	24	11	19	8	16	50	9	5	Fat Clay	CH
6	60	26	35	16	18	11	24	32	10	7	Clayey	CH
7	54	27	27	13	20	9	20	45	8	6	Clayey	CH
8	52	25	27	12	17	9	18	40	6	3	Clay Loam	CH
9	56	24	32	15	19	12	26	31	9	3	Clay Loam	CH
10	66	25	42	20	18	15	31	20	10	8	Clay Loam	CH
11	50	23	27	13	18	10	21	37	11	7	Clay Silt	CH
12	49	29	22	11	23	8	18	55	5	4	Clay Silt	CH
13	60	27	33	16	20	12	25	35	7	4	Sand Clay	CH
14	57	23	34	16	19	13	27	28	8	4	Sand Clay	CH
15	57	28	29	14	21	10	21	43	7	3	Clay Loam	CH
16	53	26	28	13	20	10	15	36	8	5	Clay Loam	CH
17	49	24	27	13	18	8	21	41	7	3	Clay Silt	CH
18	50	26	31	15	20	11	24	37	7	4	Silty Clay	CH
19	52	21	32	15	16	12	25	26	5	5	Silty Clay	CH
20	66	22	44	21	18	16	35	11	7	4	Clay Loam	CH
21	58	27	31	15	20	11	23	39	9	5	Sand Clay Loam	CH
22	53	29	24	11	19	8	16	50	10	7	Clay Silt	CH
23	60	26	35	16	18	11	24	32	8	6	Clay Loam	CH
24	54	27	27	13	20	9	20	45	6	3	Sand Clay	CH
25	52	25	27	12	17	9	18	40	7	3	Sandy Loam	CH
26	56	24	32	15	19	12	26	31	7	4	Clayey	CH
27	66	25	42	20	18	15	31	20	5	5	Silty Clay	CH
28	50	23	27	13	18	10	21	37	7	4	Clay Loam	CH
29	49	29	22	11	23	8	18	55	9	5	Clayey	CH
30	60	27	33	16	20	12	25	35	10	7	Silty Clay	CH
31	57	23	34	16	19	13	27	28	8	6	Clay Loam	CH
32	57	28	29	14	21	10	21	43	6	3	Clay Loam	CH
33	53	26	28	13	20	10	21	41	9	4	Sandy Loam	CH
34	49	24	27	13	17	9	20	39	9	6	Clayey	CH

Table 2.
Plasticity index and shear strength parameters.

expansiveness (ε_{ex}), all the tested samples were in the range between 20 and 50, and this indicated medium expansive soils which were susceptible to landslides.

From the plasticity chart (**Figure 8**), all the soil samples tested were inorganic clay of high plasticity belonging to the CH group. Such soils can easily move when saturated, leading to a high incidence of landslides.

From **Figure 8**, it is evident that soils in the study area have high plasticity as already revealed by XRD results. Such soils can easily move when saturated, leading to high incidence of landslide occurrence. The presence of many cracks, observed on the soil surface during field investigations confirms the high plasticity nature of the soil materials (**Figure 9**). The presence of such cracks on the surface is also a characteristic of Vertisols with high expansive potential.

From the double hydrometer test, it can be vividly observed that most of the samples have dispersion values greater than 30%. By implication, such soil materials are susceptible to landslide occurrence. From the soil samples examined, critical dispersion values greater than 50% were established from more than 90% of the samples (**Figure 10**). Such high dispersion values imply greater susceptibility to landslides in the region. This study established that, highly dispersive soils are particularly dominant in the surface soil layers associated with greater percentages of illite/muscovite clay minerals.

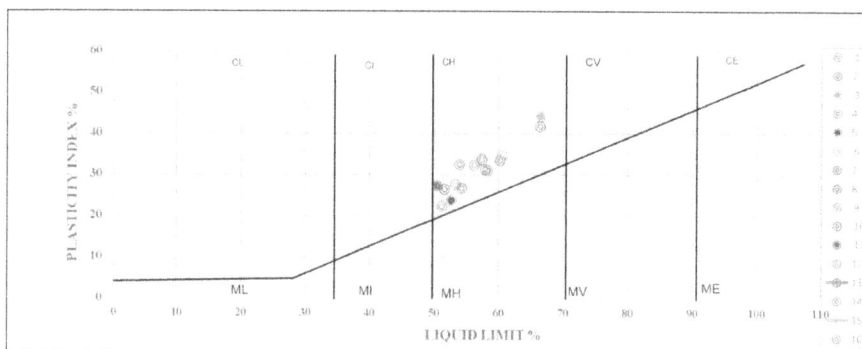

Figure 8.
The distribution of samples on the plasticity chart for the USCS.

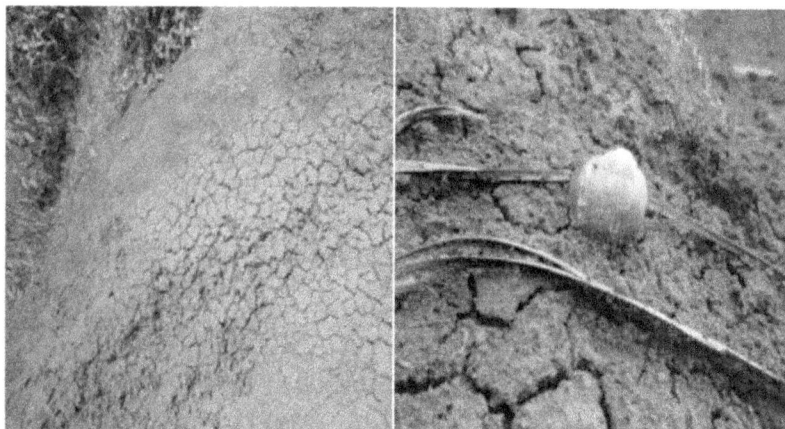

Figure 9.
Highly expansive materials with numerous surface cracks. Photo credit, Nseka, march 2019.

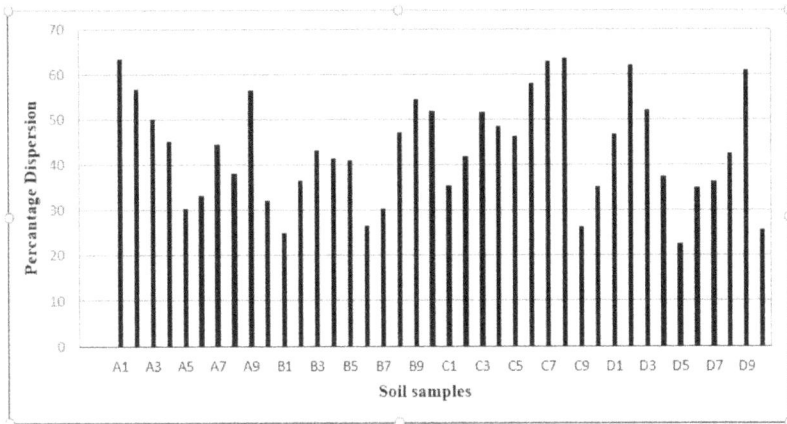

Figure 10.
Double hydrometer test results.

The linear shrinkage computed from the soil samples ranged between 10.53 and 20.76 (**Table 2**). The computed weighted plasticity index (PI_w) from the analysed soil materials ranged between 17.92% and 34.92%, averaging 28.4%.

The computed weighted plasticity index for most of the tested soils is above the 20% threshold for expansive soils, signifying highly unstable soils. All the tested samples had expansiveness (ε_{ex}) between 20 and 50 indicating medium expansive soils which are highly vulnerable to slope failure. More than 90% of the analysed samples have expansiveness above the 20% threshold. More than 80% of the samples have values between 20 and 50% showing medium expansive soils (**Table 2**).

3.5 Shear strength parameters

Shear strength parameter test results show that the soils have low cohesion (C). From the samples analysed, soil cohesion ranged between 5.2 kPa and 11.1 kPa averaging 8.2 kPa. The angle of internal friction obtained from the analysed soil samples ranged between 2.6° and 8.1° averaging 5.4° (**Table 2**). From the soils completely saturated with water, a lower cohesion of 5.2 kPa was detected. A minimal cohesion value of 8.2 kPa was considered as the critical state equilibrium all over the area. This is due to the fact that slope failure is expected to occur when soil materials are saturated. The present study established that most of tested soil samples had a very low internal friction angle (<8.5°). Such soil materials with very low internal friction angle were considered weak with high vulnerable to landslides.

3.6 Soil water infiltration

High infiltration rates were observed in the top soils with depth ranging from 0.3 to 0.8 m, but drastically reduced in the sub soils. Steady infiltration rates greatly varied with topographic characteristics and land use/cover types. Higher infiltration values were noticed in the lower slope elements than the upper sections (**Figure 11**). From the infiltration experiments conducted, it was revealed that the average soil-water infiltration was 24 cm/h^{-1} in uppermost slope sections. On the upper-middle, lower-middle slope sections and bottom valleys, soil-water infiltration rates of 30 cm/h^{-1}, 70 cm/h^{-1} and greater than 80 cm/h^{-1} respectively were achieved from the experiments conducted. Along the hollows within the upper slope sections, the

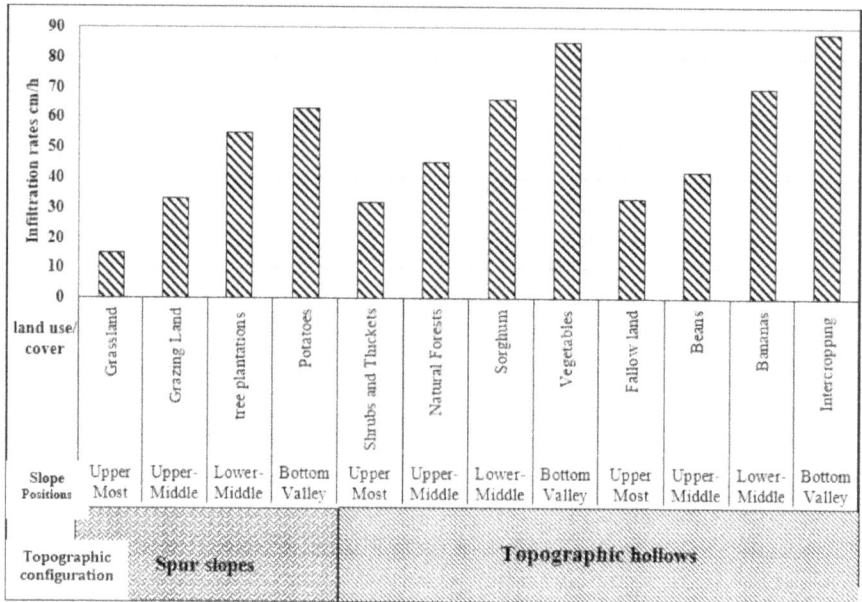

Figure 11.
Infiltration rates along slope positions, topographic configurations and land uses/covers.

observed infiltration rates were greater than 30 cm/h^{-1} and less than 12 cm/h^{-1} along the spur slope sections. Whereas the infiltration rates within topographic hollows along the middle slope sections was greater than 70 cm/h^{-1}, it was less than 45 cm/h^{-1} on the spur slope elements (**Figure 11**). Within the top soil layers, high infiltration rates are explained by the dominancy of loamy sandy soils. The predominance of clay materials in the subsoil, with claypans distinctly underlying the top soils, limits water infiltration in the study area. Such soil characteristics affect their response to incoming rainfall and consequently the timing of landslides, as explored in the discussion section.

The land use and cover characteristics influence soil water infiltration in the study area. Within the agricultural land uses particularly along cultivated zones, infiltration rates were noticed to be higher than those observed on natural land cover types. The infiltration rates along the agricultural land uses were generally greater than 65 cm/h^{-1} with the exception of beans covered areas where infiltration rates were noticed to be lower than 42 cm/h^{-1}. On the natural land cover types including grasslands, thickets and shrubs, infiltration values lower than 30 cm/h^{-1} were generally observed (**Figure 11**). Areas covered by forests, however, tended to have infiltration values higher than 45 cm/h^{-1} (**Figure 11**). Basing on these results, it was deduced that infiltration values in the study area vary greatly between rapid and very rapid. They also vary with slope characteristics including gradient and curvature as well as land use and cover properties. Following the experiments conducted, it was concluded that the steady state water infiltration values in the study area range from 12.2 cm h^{-1} to 88.5 cm h^{-1}.

3.7 Rainfall distribution and soil behaviour

The behaviour of soil materials and its susceptibility to landslide occurrence greatly depends on rainfall amounts and distribution in the region. During the

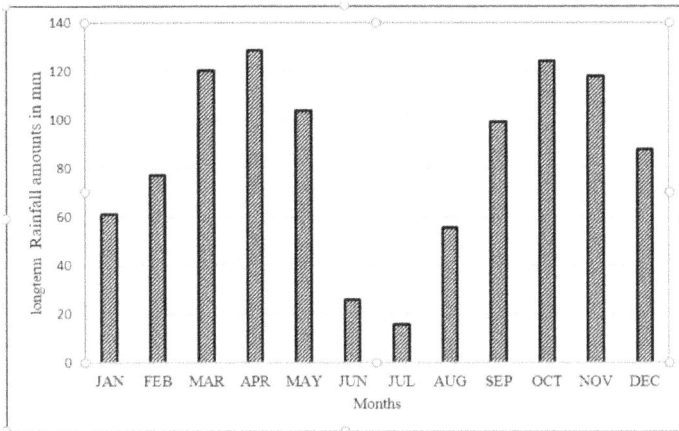

Figure 12.
Long term average monthly rainfall distribution for 1980 to 2020.

analysis of landslide occurrence, considering the influence of soil antecedent moisture and its implication on soil pore water pressure is very important. Monthly
rainfall distribution shows that March, April, October and November are the wettest months in the study area (**Figure 12**). Rainfall amounts and distribution have
great implications for soil behaviour and hence landslide occurrence in the study
area. Seasonal rainfall distribution shows more rains are received during the MAM
(352.5 mm) and SON (327.8 mm) seasons, while DJF (238.4 mm) and JJA
(104.72 mm) seasons receive less rainfall (Appendix 2). It is noteworthy, however,
that landslide occurrence in the study area is not linked to individual rainfall events,
but correspond with seasonal rainfall distribution.

Following an interaction with the local communities as well as local government
reports, it was established that most of the landslides are experienced during
months of May and November. These are, however, not the wettest months in the
region (Appendix 2). In 2010, for example, more landslides occurred during the
month of May which, however, received lower rainfall amounts (97.7 mm) compared to the preceding months of March (149 mm) and April (133 mm). During
2013, landslides were similarly experienced during the month of November with
lower rainfall amounts (122. 2 mm) compared to the preceding months of September (134.1 mm) and October (154 mm) (Appendices 2 and 3). It is therefore,
noteworthy that landslides in the study area do not necessarily occur in the wettest
months of the year. The implications of this phenomenon is unravelled in the
subsequent section.

4. Discussion

4.1 Soil profile characteristics

Deep soil profiles ranging between 2.5 and 7 m are a major characteristic of the
study area. Deeper soil profiles are more pronounced along topographic hollows and
valley bottoms. Soil depth forms one of the conditions for assessing the stability of
the soil materials and landslide susceptibility of the landscape [40]. Soil depth and
its moisture content determine how water can be stored in the soil before saturation

is reached [41]. Although most of the ridges in the study are characterised by deep to very deep soils profiles (greater than 6 m on most slopes), majority of the landslide features are shallow, concentrating within 1 to 3 m of the profile. Shallow landslide scars are not uncommon in slope sections covered by deep and very deep soil materials. As has been illustrated by [7], most sections covered by deep soils on the slopes of Mt. Elgon in Eastern Uganda experience deep seated landslides. Shallow landslides on deep soil covered slopes in the Kigezi highlands was, therefore, considered as an anomaly. Soil profile analysis was undertaken to establish the cause of this irregularity. The profile characterisation indicated the existence of 0.9 to 3 m dense claypans within the profiles. Following the infiltration experiments conducted, it was established that claypans decreased the rate of water infiltration within the soil materials. The restraint of perpendicular water flow within the soil materials by claypans has also been confirmed by [42]. The accumulating water leads to saturation of claypans sandwiched between more stable materials. Saturated claypans can act as a sliding surface for the overlying materials, consequently inducing landslides. Other studies elsewhere also confirm that variations of claypan profile properties over the landscape greatly influence soil-water holding capacity [35, 43]. It can, therefore, be inferred that the occurrence and characteristics of landslides in Kigezi highlands is highly influenced by the presence and position of claypan horizons within the soil profile.

4.2 Particle size distribution

Fine-grained silt and clay soils predominate the study area. From the mechanical analysis conducted to determine soil particle sizes, it was established that clay is the dominant material, with greater than 40%. The distribution of sand and silt in the soil materials was less than 35% and 25% respectively. The soil materials in Kigezi highlands can be classified as Vertisols due to the high clay content of more than than 35% on average and plasticity index (PI) greater than 33%. Such soils are known for inducing landslides [1]. Vertic soils characteristically expand when wet and shrink in dry conditions due to their high clay content [9]. The presence of large amounts of clay in soils of the study area is a major factor in landslide occurrence, since it affects the stability of the soils when wet. Likewise [7], also observed that the susceptibility to landslide occurrence on the slopes of Mount Elgon is due to abundance of fine-grained materials in the subsurface. The study results are, therefore, consistent with several studies elsewhere which have demonstrated the influence of high clay content on landslide occurrence [1, 18, 44].

4.3 Clay mineralogy

XRD Clay mineral analyses indicated the presence of moderately expansive clays, particularly illite/muscovite. Previous studies indicate that the presence of illite clays can lead to landslide occurrence due to their swelling potential and low shear strength [10, 45]. In the same vein [7], also confirmed that the occurrence of landslides on Mount Elgon slopes in Eastern Uganda is associated with the existence of greater amounts of kaolinite and illite clay minerals in the soil materials. The existence of significant amounts of illite/muscovite clay minerals in the study area further confirmed that Vertisols with high shrink-swell properties can result into landslides. The behaviour of the soil materials greatly determine the timing of landslide occurrence during rainfall seasons. There is fast flow of water through the surface soil materials with greater amounts of quartz into the deep soil profiles with clay abundance at the start of the rainfall season. Greater amounts of illite/muscovite clay minerals in the subsurface absorb the arriving water resulting into moisture build-up. The amassing water in the soil materials leads to soil behavioural change which can swell and loose cohesion [46]. This phenomenon, therefore, explains why landslides in the study area are not experienced at the beginning of the rainfall season or immediately after extreme rainfall events, as is the case with Mt. Elgon region in Eastern Uganda.

4.4 Soil dispersion

The study area is characterised by soil materials associated with high plasticity and are, therefore, inorganic in nature (CH), signifying weak soils of high saturation (**Figure 7**). Results from soil dispersion tests revealed high plasticity index of greater than 30% signifying Vertic soils. During continuous rainfall events, such soils with high plasticity can easily slide. The LLs for all samples analysed was above 50%, signifying high plasticity. Following soil expansiveness analysis, it was established that the average weighted plasticity index (PI_w) and expansiveness were 26.4% and 32.8% respectively. By implication, such soils are highly dispersive and excessively susceptible to landslide occurrence [10]. Landslide occurrence in the study area is, therefore, associated with expansive soils which shrink and swell leading to loss of soil strength. The expansive potential of the soils influenced by high clay content and type especially illite/muscovite is one of the major factors promoting landslides in Kigezi highlands.

4.5 Soil water infiltration

Soil water infiltration was noted to vary across slope position and topographic configuration. Whereas upper slope sections and spur slopes with shallow soils experience low infiltration rates, lower slope sections and topographic hollows with deeper soils are associated with high infiltration rates. This variation in infiltration rates also signifies differences in soil saturation levels ("For example, see [47]"). The lower slope sections are associated with greater saturation rates which result into saturation overland flow processes. Along the topographic slope elements, the saturation overland flow incrementally moves upslope from the slope base [48]. Materials along topographic hollows remain saturated most times due to greater clay content dominated by illite/muscovite minerals. This phenomenon leads to decreased soil material strength along the topographic hollows, inducing landslide occurrence. Equally noticed in this study was a relationship between topographic characteristics and soil water infiltration. Greater soil-water infiltration values are experienced along topographic hollows and lower slope elements than the spurs and upper sections. Such variations in infiltration values in relation to topographic characteristics are associated with differences in soil depth along the slope configurations. This study, therefore, approves that using the soil-water infiltration experiments along different slope sections and gradients it is possible to predict landslide occurrence.

4.6 Change in soil behaviour with rainfall distribution

As opposed to the situation on the slopes of Mt. Elgon in Eastern Uganda, landslide occurrence in Kigezi highlands is not related to extreme rainfall events. On the slopes of Mt. Elgon in Eastern Uganda, majority of the landslides are commonly witnessed during or immediately after extreme rainfall events [1, 7, 18]. Local communities and regional government reports confirmed that landslide occurrence in the study area is usually not experienced during or immediately after peak rainfall seasons. Paradoxically, they occur during the less wet months of the rainfall season. Most landslides in the study area are experienced during the months of May and November, despite the preceding months of April and October receiving more rainfall amounts. This phenomenon can be explained by the unique infiltration dynamics through quartz dominated top soil layers and saturation of the claypans dominated by illite/muscovite clays in the lower soil horizons. This leads to antecedent moisture building up in the sub soil materials as more rainfall is received hence landslide occurrence in the region. Antecedent soil moisture condition prior to a rainfall event has been confirmed as the most significant factor in landslide occurrence [49].

5. Conclusion

Deep soil profiles are a major characteristic of the study area. Notwithstanding the deep soil profiles on most slope elements in the study area, majority of the landslide scars are shallow, occurring within less than half of the profile due to presence of claypans. The claypans act as a slipping zone for the overlying soil materials. The study area is dominated by fine silt and clay soil materials. In association with greater amounts of clay percentage of more than 35% on average and PI greater than 33%, the soil materials in the study area are classified as Vertisols, which are synonymous with landslide occurrence. The predominance of reasonable amounts of expansive clays, mainly illite in the study area influences the stability and vulnerability of slope materials to landslide occurrence. In Kigezi highlands, landslides are not normally experienced during or immediately after extreme rainfall events but occur later in the rainfall season due to initial infiltration through quartz dominated upper soil layers, before illite/muscovite clays in the lower soil horizons get saturated. This behavioural change in the soil material due to moisture content is, therefore, the major trigger of landslides in Kigezi highlands. An understanding of these soil characteristics is an important step in landslide hazard mitigation in Kigezi highlands.

Acknowledgements

The authors gratefully acknowledge the research grant from Makerere University – Swedish International Development Cooperation Agency (SIDA) Phase IV (2015/2020 Agreement) – Building Resilient Ecosystems and livelihoods to Climate Change and Disaster Risk (BREAD) project 331 research component, which funded travel and fieldwork for this study.

Conflict of interest

The authors have no conflict of interest.

Appendices

Appendix 1. Soil morphological characteristics

Landslide site	Soil profile	Gradient (°)	Soil depth (m)	Location of clay pan	Horizon A (m)	Percentage distribution			Horizon B (m)	Percentage distribution			Horizon C (m)	Percentage distribution		
						Sand	Silt	Clay		Sand	Silt	Clay		Sand	Silt	Clay
1	P1	38	0.92	0.55	0.61	44	33	23	0.82	28	24	48	0.88	12	28	60
	P2	31	1.72	1.02	0.75	44	32	25	1.21	27	25	48	1.55	18	31	51
	P3	21	3.32	1.17	1.17	54	23	23	1.94	21	23	56	2.94	23	27	50
	P4	9	5.11	3.14	1.44	38	34	28	2.82	22	16	62	4.12	21	26	53
2	P5	47	0.71	0.43	0.67	40	31	29	0.69	23	24	53	0.7	20	25	55
	P6	26	2.12	1.22	0.82	33	40	27	1.44	29	25	46	2.02	18	19	63
	P7	18	3.83	2.16	0.93	38	32	30	2.11	30	24	46	3.21	22	26	52
	P8	10	6.21	3.11	1.11	42	30	28	3.21	21	17	62	4.23	17	22	61
3	P9	34	0.62	0.72	0.62	39	36	25	0.92	24	21	55	1.22	23	26	51
	P10	28	2.03	1.44	0.74	42	31	27	1.23	28	25	47	1.95	21	26	53
	P11	21	3.92	2.82	0.92	35	37	28	1.94	19	33	48	3.32	19	27	54
	P12	11	5.82	3.32	1.23	40	34	26	3.11	24	20	56	4.74	21	30	49
4	P13	48	0.78	0.69	0.65	43	36	21	0.75	28	27	45	0.91	23	27	50
	P14	30	1.83	1.12	0.82	44	30	26	1.22	22	14	64	1.73	19	37	44
	P15	16	3.41	1.92	0.93	43	32	25	1.88	21	29	50	3.11	22	23	55
	P16	8	6.34	3.17	1.13	40	34	26	3.11	31	18	51	5.44	20	27	53
5	P17	43	0.65	0.73	0.61	41	33	26	0.82	25	28	47	1.11	21	22	57
	P18	31	1.74	0.93	0.73	44	32	24	1.44	20	28	52	1.72	22	25	53
	P19	19	3.12	1.83	0.84	45	32	23	2.11	24	26	50	2.88	18	28	54
	P20	7	6.81	3.24	1.06	47	30	23	3.33	26	30	44	5.81	21	25	54

Landslide site	Soil profile	Gradient (°)	Soil depth (m)	Location of clay pan	Horizon A (m)	Percentage distribution			Horizon B (m)	Percentage distribution			Horizon C (m)	Percentage distribution		
						Sand	Silt	Clay		Sand	Silt	Clay		Sand	Silt	Clay
6	P21	46	0.87	0.61	0.66	44	31	25	0.93	23	30	47	1.31	20	27	53
	P22	29	2.31	1.22	0.81	50	32	18	1.71	25	27	48	2.11	19	20	61
	P23	15	4.12	2.12	1.22	42	22	36	2.34	22	36	42	3.88	19	31	50
	P24	11	7.12	3.11	1.54	40	28	32	3.33	21	18	61	4.78	22	29	49
7	P25	44	0.93	0.82	0.77	52	32	16	0.87	27	23	50	0.93	23	19	58
	P26	33	2.21	1.36	0.88	43	29	28	1.32	26	24	50	2.01	16	34	50
	P27	18	3.36	1.97	0.91	46	36	18	2.11	22	31	47	3.12	23	25	52
	P28	13	6.61	3.92	1.22	55	35	10	3.42	18	36	46	4.68	18	36	46
8	P29	42	0.96	0.78	0.71	50	35	15	0.88	31	23	46	0.97	21	26	53
	P30	25	2.42	1.44	0.92	44	37	19	1.94	28	23	49	2.11	22	28	50
	P31	20	4.21	2.33	0.95	48	30	22	2.88	30	25	45	3.89	18	21	61
	P32	12	7.21	3.12	1.23	47	37	16	3.44	22	36	42	5.88	20	27	53
9	P33	48	0.83	0.48	0.66	49	33	18	0.84	30	21	49	0.95	23	26	51
	P34	28	1.94	0.78	0.77	42	36	28	0.91	28	21	51	1.44	19	27	54
	P35	22	3.56	2.21	0.93	53	31	16	2.45	27	25	48	3.22	18	30	52
	P36	8	6.72	3.37	1.11	50	27	23	3.11	19	37	44	5.77	19	24	57
10	P37	45	0.94	0.53	0.71	44	33	23	0.88	29	23	48	1.17	21	24	55
	P38	34	2.44	1.52	0.93	40	33	27	1.98	23	27	50	2.23	23	20	57
	P39	17	4.23	2.17	1.03	48	33	19	2.44	17	20	63	4.11	22	21	57
	P40	11	6.83	3.13	1.33	45	30	25	3.12	18	35	47	5.88	22	18	60

Appendix 2. Temporal distribution of rainfall in Kabale highlands

Year	Jan	Feb	Mar	Apr	May	Jun	Jul	Aug	Sep	Oct	Nov	Dec
1980	56.6	75.3	51.8	164	188	5.3	8.2	39.4	103.3	111.4	137.5	47.8
1981	67	16.7	147.8	138.2	102.8	22.4	7.3	128.8	58.2	122.4	58	74.7
1982	18.9	8.3	42.7	272.2	131.9	15.5	7.5	2.8	107.6	123.4	167.5	22.3
1983	13.4	51.8	88.7	126.3	45.2	12.5	19.2	88.4	70.4	236.5	74.3	95
1984	37.2	108.3	175.4	157.3	20.3	11.2	49.7	21.6	78.8	113.7	97.2	128.2
1985	32.4	34.7	103.4	148.5	40.6	2.8	12.2	23.8	90.4	122.4	72.5	37
1986	0	0	102.6	184	64.3	31.7	18.2	16.4	44.3	121.7	71.1	488.4
1987	82.7	99.2	113.5	131.1	178.4	44.6	3.9	23.3	108	129.1	261.5	31.3
1988	83.9	90.9	161.7	139.1	82.2	4	64.2	142.6	166	132.8	60.8	47
1989	34.7	99.8	82	89.6	134.9	35.6	10.1	74.7	180.8	128.5	72.3	74.3
1990	49	161.8	128.4	182.6	72.6	0	0	45	158.5	66.5	93.6	68.9
1991	72	73	158	101	117	39	18	18	54	135	51	71
1992	18	49	151	98	50	54	27	16	149	205	89	74
1993	96	28	176	87	160	34	0	61	10	59	95	77
1994	58	70	128	150	87	2	2	79	148	125	134	93
1995	45	128	105	82	147	123	1	6	114	166	105	102
1996	70	56	146	93	46	72	50	118	123	144	202	102
1997	101	0	114	122	149	33	27	37	25	155	196	151
1998	184	97	101	171	170	19	25	23	87	154	58	80
1999	77	37	145	72	51	0	0	167	65	87	116	49
2000	50.9	83.5	118.9	120	55.7	8.1	5.9	69.6	69.9	179.8	146.8	83.8
2001	86.3	51.2	83.9	135.7	77.8	22.4	46.7	65.7	231.1	201.9	139.5	63.9
2002	120.2	89.7	63.1	74.9	115.7	0	4.4	48.4	49.5	187.6	91.5	91
2003	66.9	80.6	74.7	139.1	96	29	22.8	25.5	82.6	86.5	94	57.3
2004	69.4	93.8	84.5	183.2	84.6	0	1.1	31.9	148.8	76.9	114.2	124.9
2005	25.7	121.8	170.1	123.4	122.1	40.5	0	29	84	107.6	66.1	41.1
2006	85.5	133.5	127.7	112.7	207.8	2.9	30.1	79.5	74.2	70.7	156.2	62.3
2007	55.2	102.5	80.3	103.6	87.9	34.1	42	23.6	99.5	112.1	162.9	25.2
2008	99	65.3	206.1	54.2	53.5	65.9	24	36.5	77.4	172.8	107.6	99.1
2009	61	114.2	122.7	99.5	90.7	19.8	1.1	94.6	87	86.6	174.1	98.8
2010	97.7	189	149.1	132.9	97.7	9.3	1.4	16.4	124.1	197	86.4	71
2011	33.5	64	139.2	88.2	63.9	62.5	12.1	103.8	71.7	73.9	157.5	54.7
2012	2.8	52.8	110.6	197	146.8	11.6	9.4	62.6	95.7	114.4	179.8	128
2013	30.1	101.1	142.2	98	192	21.1	3.2	74.5	134	154	122.2	77.1

Source: Kabale meteorology station, weather data: WMO No. 63726, National No. 91290000, station name KABALE, Elevation 1867 m, Latitude 01°15', Longitude 29°59'.

Appendix 3. Relationship between seasonal rainfall distribution and landslide occurrence

Year	Season	Rainfall in mm	No of landslide occurrences	Month of occurrence
2005	MAM	414.6	6	May
2006	MAM	447.2	5	May
2007	MAM	272.8	5	April
2008	MAM	464.8	13	May
2009	SON	348.7	5	October
2010	MAM	457.7	31	May
2011	MAM	461.2	22	May
2012	MAM	453.4	8	May
2013	SON	304.1	5	November
2014	SON	253.2	3	October
2015	MAM	254.7	17	May
2016	SON	298.3	9	October
2017	MAM	213.8	11	May
2018	MAM	417.8	21	April
2019	SON	314.7	13	November
2020	MAM	339.9	17	May

Author details

Denis Nseka[1*], Vincent Kakembio[2], Frank Mugagga[1], Henry Semakula[1], Hosea Opedes[1], Hannington Wasswa[1] and Patience Ayesiga[3]

1 Department of Geography, Geo-Informatics and Climatic Sciences, School of Forestry, Environmental and Geographical Sciences, Makerere University, Kampala, Uganda

2 Department of Geosciences, School of Environmental Studies, Nelson Mandela University, Port Elizabeth, South Africa

3 Department of Geography, Faculty of Education, Bishop Stuart University, Mbarara, Uganda

*Address all correspondence to: denisnseka1@gmail.com

IntechOpen

References

[1] Mugagga, F. V., Kakembo, V., Buyinza, M., 2011. A characterisation of the physical properties of soil and the implications for landslide occurrence on the slopes of mount Elgon, eastern Uganda. Journal of the International Society for the Prevention and Mitigation of natural hazards ISSN 0921-030X. DOI:10.1007/s11069-011-9896-3.

[2] Kirschbaum, D. B. and Zhou, Y. 2015. Spatial and temporal analysis of a global landslide catalog. Geomorphology DOI: 10.1016/Journal of Geomorphology. 2015.03.016.

[3] Bamutaze, Y. (2019). Morphometric Conditions Underpinning the Spatial and Temporal Dynamics of Landslide Hazards on the Volcanics of Mt. Elgon, Eastern Uganda. In Emmerging Voices on Natural Hazards. Elsevier Inc.

[4] López-Davalillo, B., Monod, M. I., Alvarez-Fernandez, G., Herrera Garcia, J., Darrozes, C., Gonzalez-Nicieza, Olivier, M. 2014. Morphology and causes of landslides in Portalet area (Spanish Pyrenees): Probabilistic analysis by means of numerical modelling. Engineering Failure Analysis 36.

[5] Hong, Y., Adler, R., Huffman, G., 2007. Use of satellite remote sensing data in the mapping of global landslide susceptibility. Natural Hazards 43, 245-256.

[6] Vagen, T.G., 2010. Africa soil information service: Hydrologically corrected/adjusted SRTM DEM (AfrHySRTM). Nairobi, Kenya and Palisades, NY: International Centre for Tropical Agriculture – Tropical Soil Biology and Fertility Institute (CIAT-TSBF), World Agroforestry Centre (ICRAF), Centre for International Earth Science Information Network (CIESIN), Columbia University.

[7] Kitutu, M. G., Muwanga, A., Poesen, J., Deckers, J. A. 2009. Influence of soil properties on landslide occurrence in Bududa district, eastern Uganda. African Journal of Agriculture Res 4(7), 611–620.

[8] Zinck, J.A., 2013. Geopedology: Elements of geomorphology for soil and geo-hazard studies. ITC Special Lecture Notes Series: ISBN: 978-90-6164-352-4, 13.

[9] Mukasa-Tebandeke, I. Z., Ssebuwufu, P. J. M., Nyanzi, S. A., Schumann, A., Nyakairu, G. W. A., Ntale, M., and Lugolobi, F., 2015. The elemental, mineralogical, IR, DTA and XRD analyses characterized clays and clay minerals of central and eastern Uganda. Advances in Materials Physics and Chemistry 5, 67-86.

[10] Yalcin, A. 2011. A geotechnical study on the landslides in the Trabzon Province, NE, Turkey. Applied Clay Science 42, 11–19.

[11] Fauziah, A., Yahaya, A, S. and Farooqi, M. A. 2006. Characterization and geotechnical properties of Penang residual soils with emphasis on landslides. American Journal of Environmental Sciences 2, 121-128.

[12] Bell, F.G. and Walker, D.J.H. 2000. A further examination of the nature of dispersive soils in Natal, South Africa. Quarterly Journal of Engineering Geology and Hydrogeology, 33, 197–199.

[13] Bell, F.G. 2004. Engineering Geology and Construction. CRC Press, Taylor and Francis. ISBN; 0415259398, 9780415259392

[14] Baynes, F.J. 2008. Anticipating problem soils on linear projects. Conference proceedings on problem soils in South Africa, 3-4 November 2008, 9-21.

[15] Williams, A.A.B., Pidgeon, J.T. and Day, P.W. 1985. Expansive soils. Transactions of the South African Institution of Civil Engineers, 27, 367-397.

[16] Zung, A.B., Sorensen, C.J., Winthers, E., 2009. Landslide soils and geomorphology in Bridger/Teton Forest Northwest Wyoming. Physical Geography, 30(6), 501–516.

[17] Roller, S., Wittmann, H., Kastowski, M., Hinderer, M., 2012. Erosion of the Rwenzori Mountains, east African rift, from in situ-produced cosmogenic 10Be. Journal of Geophysical Research, 117, F03003.

[18] Knapen, A., Kitutu, M.G., Poesen, J., Breugelmans, W., Deckers, J., Muwanga, A., 2006. Landslides in a densely populated county at the footsteps of mount Elgon (Uganda): Characteristics and causal factors. Geomorphology, 73, 149–165.

[19] Kitutu, M.G., Poesen, J.M., Deckers, J., 2011. Farmer's perception on landslide occurrences in Bududa District, eastern Uganda. African Journal of Agricultural Research, 6, 7-18.

[20] UBOS (Uganda Bureau of Statistics). 2017. Statistical Abstracts 2017. Ministry of Finance, Planning and Economic Development, Uganda. http://www.ubos.org

[21] Mugagga, F., N. Nakanjakko., B. Nakileza., D. Nseka 2020. Vulnerability of smallholder sorghum farmers to climate variability in a heterogeneous landscape of Kigezi highlands, South-Western Uganda. Journal of Disaster Risk Studies (Jàmbá), 12(1), a849. DOI: 10.4102/jamba.v12i1.849. https://jamba. org.za/index.php/jamba/article/view/849

[22] Nseka, D., Mugagga, F., Opedes, H., Ayesiga, P., Wasswa, H., Mugume, I., Nimusiima, A., & Nalwanga, F. 2021. The damage caused by landslides in socio-economic spheres within the Kigezi highlands of South Western Uganda. Journal of environmental and socio-economic studies, DOI:10.2478/environ-2021-0003, 9, 1: 23-34.

[23] Bagoora, F.D.K., 1993. An assessment of some causes and effects of soil erosion hazard in Kabale Highland, South Western Uganda, and people's attitude towards conservation. In Abdellatif (ed.) resource use and conservation: Faculty of Social Sciences; Mohammed V. university, Rabat Morocco. Mountain Research and Development, Vol 8.

[24] Nseka, D., Kakembo, V., Bamutaze, Y. and Mugagga, F. 2019. Analysis of topographic parameters underpinning landslide occurrence in Kigezi highlands of South Western Uganda, Journal of Natural Hazards. Springer, DOI: 10.1007/s11069-019-03787-x

[25] Bagoora, F.D.K., 1989. A preliminary investigation into the consequences of inadequate conservation policies on steep slopes of the Rukiga highlands, South Western Uganda, In Thomas D. B, E. K Biamah, A. M Kilewe (eds.); soil conservation in Kenya, Dept. of Agri. Er, Univ. of Nairobi, Kenya.

[26] Bagoora FDK (1988) Soil erosion, mass wasting risk in the highland areas of Uganda: Mountain Research and Development, Vol 8.

[27] NEMA (National Environment Management Authority). 2018. State of Environment Report for Uganda for 2017/18. National Environment Management Authority, Kampala, Uganda. http://www.nemaug.org

[28] UBOS (Uganda Bureau of Statistics). 2016. Statistical Abstracts 2016. Ministry of Finance, Planning and Economic Development, Uganda. http://www.ubos.org

[29] NEMA (National Environment Management Authority). 2017. State of Environment Report for Uganda for 2016/17. National Environment Management Authority, Kampala, Uganda. http://www.nemaug.org

[30] NEMA (National Environment Management Authority). 2016. State of Environment Report for Uganda for 2015/16. National Environment Management Authority, Kampala, Uganda. http://www.nemaug.org

[31] NEMA (National Environment Management Authority). 2020. State of Environment Report for Uganda for 2019/20. National Environment Management Authority, Kampala, Uganda. http://www.nemaug.org

[32] FAO., 2006. Food and Agriculture Organization of the United Nations: Guidelines for soil description. Rome (Italy). ISBN: 92-5-105521-1.

[33] ASTM D3385., 2003. Standard test method for infiltration rate of soils in field using double-ring infiltrometer. Annual book of ASTM standards, vol. 4.02. ASTM International.

[34] Philips, C. E. and Kitch. W.A., 2011. A review of methods for characterization of site infiltration with design recommendations. Paper presented at 43rd Symposium on Engineering Geology and Geotechnical Engineering, University of Las Vegas, Las Vegas, NV, March 23-25, 2011.

[35] Ahmed, F., Gulliver, J.S. and Nieber, J.L., 2011. A new technique to measure infiltration rate for assessing infiltration of BMPs. Presented at 12th international conference on urban drainage, Porto Alegre/Brazil, 11-16 September 2011.

[36] British Standards Institution. British Standards 1377: 1995 and British Standard 1377: 1990. Methods of test for soils for civil engineering purposes, London, 1990.

[37] Moussadek, R., Laghrour, M., Mrabet, R., Van Ranst, E., Badraoui, M., Mekkaoui, M., 2017. Morocco's Vertisols characterization. Journal of Materials and Environmental Sciences ISSN; 2028–2508, Volume 8, Issue 11, pp. 3932-3942.

[38] Gao, J. and Maro, J., 2010. Topographic controls on evolution of shallow landslides in pastoral Wairarapa, New Zealand, 1979-2003. Geomorphology, 114(3), 373–381. DOI: 10.1016/j.geomorph.2009.08.002

[39] Okalebo, J.R., Gathua, K.W., Woomer, P.L., 1993. Laboratory Methods of Soil and Plant Analysis: A Working Manual, Soil Science Society of East Africa, EPZ (Kenya) Limited, Nairobi, Kenya.

[40] Liang, W. L. and Uchida, T., 2014. Effects of topography and soil depth on saturated-zone dynamics in steep hillslopes explored using the three-dimensional Richards' equation. Journal of Hydrology, 510, 124–136. DOI: 10.1016/j.jhydrol.2013.12.029.

[41] Merino-Martín, L., Moreno-de las Heras, M., Espigares, T. and Nicolau, J. M., 2015. Overland flow directs soil moisture and ecosystem processes at patch scale in Mediterranean restored hillslopes. Catena, 133, 71–84. DOI: 10.1016/j.catena.2015.05.002

[42] Jiang, F.S., Huang, Y.H., Wang, M. K., Lin, J.S., Zhao, G., Ge, H.L., 2014. Effects of Rainfall Intensity and Slope Gradient on Steep Colluvial Deposit Erosion in Southeast China. Soil Science. Society of American Journal.

[43] Xiang, W., Cui, D., Liu, L., 2007. Experimental study on sliding soil of ionic soil stabilizer-reinforces. Earth Sciences 32(3), 397–402.

[44] Wati, S.E., Hastuti, T., Wijojo, S. and Pinem, F., 2010. Landslide susceptibility mapping with heuristic

approach in mountainous area. A case study in Tawangmangu sub district, Central Java, Indonesia. International achieves of the photogrammetry, Remote Sensing and Spatial Information Science, 38, 248–253.

[45] Yalcin, A., 2007. The effects of clay on landslides: A case study. Applied Clay Science, 38, 78–85.

[46] Van Den Eeckhaut, M., Poesen, J., Hervas, J., 2013. Mass-movement causes: Overloading. In: Schroder, J.F. (Ed.). Treatise on Geomorphology, Vol 7, Mountain and Hillslope Geomorphology. Academic, San Diego, 200–206

[47] Rousseau, M., Cerdan, O., Ern, A., Le Maître, O. and Sochala, P., 2012. Study of overland flow with uncertain infiltration using stochastic tools. Advances in Water Resources, 38, 1–12. DOI:10.1016/j.advwatres.2011.12.004

[48] Reynolds, W.D., Elrick, D.E., Youngs, E.G., Amoozegar, A., Booltink, H.W.G. and Bouma, J., 2002. Saturated and field-saturated water flow parameters. In Dane J. H and G.C. Topp (Eds.) methods of soil analysis. Part 4. Physical methods. SSSA book Ser. 5. SSSA, Madison, WI, 797–878.

[49] Guan-Wei, L. and Hongey, C., 2012. The relationship of rainfall energy with landslides and sediment delivery. Engineering Geology, 125, 108-118.

Chapter 4

Landslide Mitigation through Biocementation

Azizul Moqsud

Abstract

Landslide and other geo-disasters are causing a great damage to people and the resources all over the world. An environment friendly countermeasure of landslide disasters is necessary. Microbially induced calcite precipitation (MICP) is a bio-cementation process that can improve the geotechnical properties of granular soils through the precipitation of calcium carbonate (calcite) at soil particle contacts. This MICP can be an environment friendly solution for the biocementation of soil. In this study, an evaluation of biocemented soil has been carried out through direct shear test and direct simple shear test. Scanning Electron Microscopy (SEM) and Energy Dispersive X-ray Spectrometry (EDS) and X-ray Computed Tomography (X-ray CT) tests were conducted to analysis the calcite precipitation inside the biotreated soil by bacteria by using Toyoura sand and silica sand no. 4. It was observed that the amount of calcite generated in silica sand was larger than Toyoura sand. The particle shape influences the result of calcite precipitation and consequent strength of the bio-cemented sand. The amount of strength which was obtained by direct shear test and direct simple shear test indicated the granular soil became bio-stabilized within 7 days of application of nutrients from the surface. However, the amount of generated calcite was not uniformed in different layers while applying the nutrients and bacterial from the surface which was revealed by X-ray CT scan test.

Keywords: Biocementation, landslide, Microbial induced calcite precipitation

1. Introduction

Landslide and slope failures are very dangerous and caused a lot of damages to the people all over the world every year. Environmentally friendly approach to improve the soil condition is necessary for the sustainable global environment. The traditional methods to protect the land against the geo-disasters such as landslide and liquefaction are mainly mechanical or chemical approach to soil and are not environment friendly. Nature has provided a significant biologically based solution to some of the challenges that vex geotechnical infrastructure systems. Recent studies on applications of bio-mediated soil improvement methods have proved the viability of the approach for effective performance and environmental sustainability. The potential outcomes of these studies have shown greater promise of exploring a wider application of the technique in geotechnical engineering. The great promise of the use of biological treatments has been demonstrated in many applications, such as improving the shear strength and decreasing the permeability of soils [1–5] improvement in strength and durability of concrete and morter, remediation of cracks in buildings [6–10]; improvement in engineering properties of soil and cementation of sand column [11–16]. However, the uniformity of biocementation inside the soil is not well-known yet.

The objective of this research is to look at the difference of mechanism of biocementation between the Toyoura sand and silica sand no. 4 through scanning electron microscope and x-ray CT analysis.

2. Materials and methods

2.1 Microbial preparations

There are several ideas of ground improvement method utilizing microbial metabolism, among which the calcium carbonate method has been vigorously researched in Japan and abroad recently due to the applicability to the real ground and the formation of solid matter derived from microorganisms. (ATCC 11859) used in this study as a source of microorganism [16, 18]. The characteristic of this microorganism is that it has the function of decomposing urea called urease enzyme. In addition, it is known that this microorganism has pressure, temperature, salt tolerance and alkali resistance, and it has a relatively strong resistance under various ground environments. The chemical reaction at that time is shown below. Cementation action between particles is caused by calcium carbonate precipitated between the soil particles, of the applied ground.

2.2 Sand used in the experiment

Two types of sands were used in the experiment to compare the effects of the size of the particles on biocementation. Toyoura sand and silica sand no. 4 were used for biocementation. The grain size analysis of those sands is shown in **Figure 1**. The particle size of silica sand is larger than Toyoura sand.

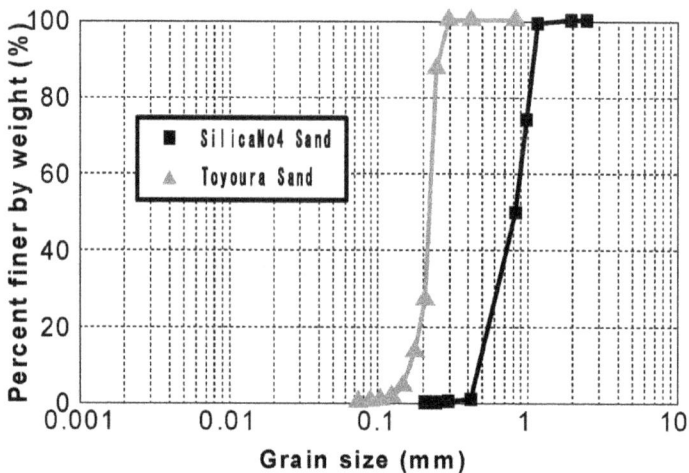

Figure 1.
Grain size analysis of Toyoura sand and silica sand used in the experiment.

2.3 Direct simple shear test

The shear strength properties measured by using direct simple shear test apparatus. The sample dimension used was 60 mm x 22.6 mm. The samples have been sheared by above mentioned apparatus under three normal loads magnitudes, namely10, 30 and 50 kN/m^2. The samples under the same load magnitude have been sheared at least three times. The constant velocity of magnitude 0.2 mm/min was applied. The test is finished when the shear strain reaches 26% (**Figure 2**).

2.4 SEM and EDS analysis

The scanning electron microscope (SEM) analysis was carried out by using JSM-7600F and consequently analyzed the energy dispersive spectroscopy (EDS) to observe the surface of the bio-cemented soil particles and the mineral amount in different samples.

2.5 Micro-focus X-ray CT system

X-ray CT scan was carried out of the treated samples after 1 week of treatment for both Toyoura sand and silica sand to observe the location of the calcite generation.

Figure 2.
Direct simple shear test apparatus.

3. Results and discussion

Figure 3 illustrates that the result if the X-ray CT scan of the 1 week bio-treated sand samples. It was observed that in silica sand the amount of CaCO3 was more than that of Toyoura sand. The shape of the sand particles has influence to generate the amount of calcite [17–22]. Another thing was observed that the amount of calcite was more in the lower portion than the upper portion of the samples. The bacteria and the nutrient was applied from the surface of the sample and this has made the influence to precipitate the calcite at the lower portion more as liquid flows through the pore spaces.

Figure 4 shows that the scanning electron microscopic view of the biotreated sand after 1 week of treatment. It was observed that the shape of the crystal is different in Toyoura and Silica sand. This type of shape of crystal might be give some influence on the strength of the biotreated sand.

Toyoura sand   Silica sand no. 4

Figure 3.
X-ray CT analysis (left photo Toyoura sand and right photo silica sand).

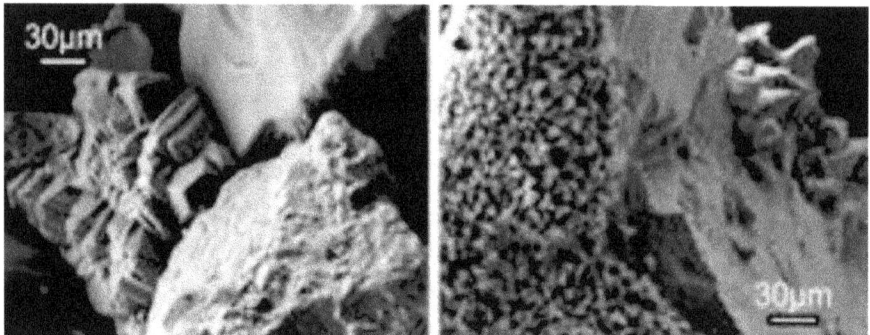

Figure 4.
Scanning electron microscope (SEM) analysis of the Toyoura (left) and silica (right)sand.

Figure 5 displays the EDS analysis of the biotreated soil samples. Rhombohedral crystal (calcite) was present on the surface of the particle and Ca element was extracted from the element mapping, and the tendency that calcium carbonate is widely distributed on the surface of the particle of silica sand than Toyoura sand.

Figure 6 shows the relationship between Pca and depth of Toyoura sand and silica sand. As shown in **Figure 1**, the particle size was larger than Toyoura sand, and it was considered that calcium carbonate covered and enlarged around the particle. Rhombohedral crystal (calcite) was present on the surface of the particle and Ca element was extracted from the element mapping, and the tendency that calcium carbonate is widely distributed on the surface of the particle of silica sand than Toyoura sand. An aggregate of rhombohedral crystals with a grain diameter of rhombohedral crystal (calcite) of 10 μm to 50 μm was observed. A spherical crystal (vaterite) different from rhombohedral crystal was observed in Toyoura sand. In silica sand, existence of spherical crystal (vaterite) could not be confirmed. In addition, it was considered that the transition from vaterite to calcite was proceeding in silica sand. As the Toyoura sand and the silica sand, the shape of the particle is different they also influenced the shape of crystals.

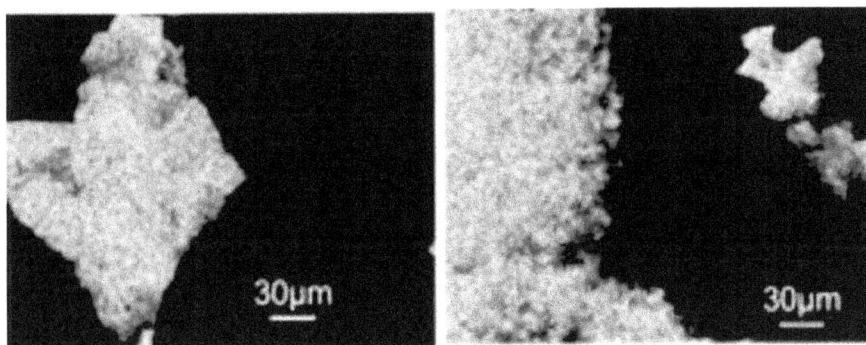

Figure 5.
EDS analysis of Toyoura (left) and silica (right) sand.

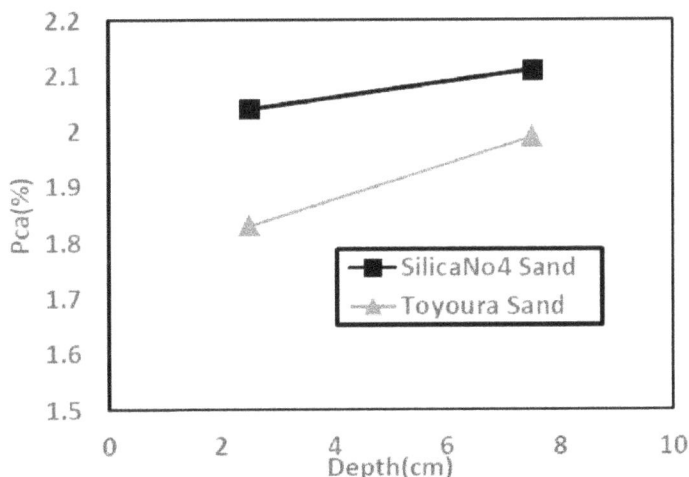

Figure 6.
Relation between calcium production and depth.

From **Figure 6** it was confirmed that silica sand Ca (calcium) element increased as compared with Toyoura sand under the same treatment condition. Since the particle size of silica is larger than Toyoura sand, calcium carbonate covers the particle surface particles, and the amount of Ca (calcium) element is higher than Toyoura sand. It was considered that it increased shear test by calcium carbonate method.

Figures 7 and **8** show the relation among the shear stress and shear strain and volumetric strain by direct simple shear test of Toyoura sand and silica sand, respectively. It shows the stress–strain relation of Toyoura sand the stress–strain relation of Toyoura sand cemented 1 Week by bio-treatment. It was observed that the shear stress value increased by 1.3 to 2.0 times in the case of no addition of microorganisms of silica sand and Toyoura sand and addition of microorganisms with 1 Week. An increase in shear stress was confirmed and from the point of comparison the shear stress of Toyoura sand was higher than the silica sand.

Figure 9 shows the relationship between the shear strength and the normal stress determined from **Figures 7** and **8**. **Table 1** shows the cohesion force

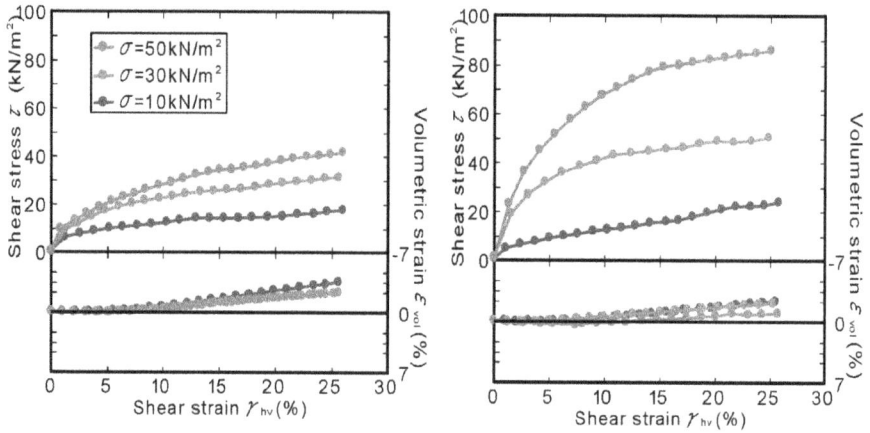

Figure 7.
Relation between shear stress and shear strain and volumetric strain of Toyoura sand without (left) and after 1 week of treatment.

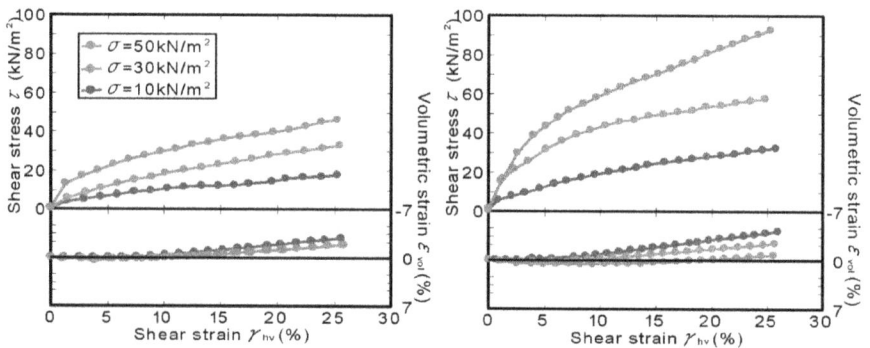

Figure 8.
Relation between shear stress and shear strain and volumetric strain of silica sand without (left) and after 1 week of treatment.

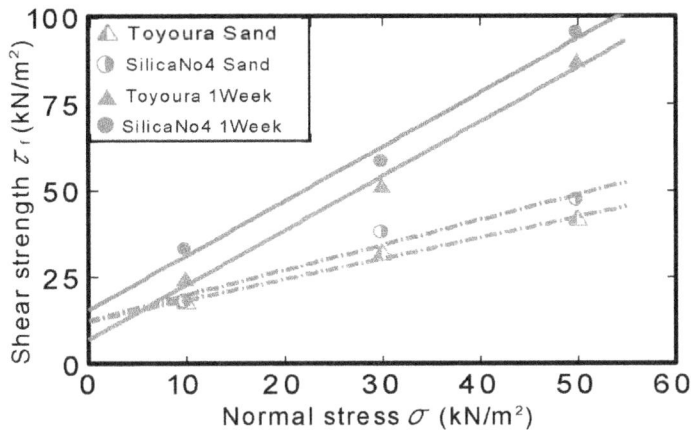

Figure 9.
Relation between shear stress and normal stress.

	Silica No 4 Sand	Toyoura Sand	Silica No 4 1 Week	Toyoura 1 Week
ϕ (°)	35.9	30.9	57.2	57.4
c (kN/m²)	12.38	12.19	15.07	6.56

Table 1.
Cohesion and degree of internal friction of Toyoura and silica sand without and 1 week of treatment.

Stress	Pca (%)	
	Silica No 4 1 Week	Toyoura 1 Week
10 (kN/m²)	2.13	1.99
30 (kN/m²)	2.06	1.84
50 (kN/m²)	2.37	2.04

Table 2.
Amount of calcite generation after 1 week.

c (kN / m2) and internal friction angle φ (°) under each condition. From
Figure 9, it was confirmed that the silica sand has increased cohesion force
c (kN/m2) and internal friction angle φ (°) because precipitation of calcium
carbonate on the particle surface increases the frictional force of the particle
surface and the increase in density due to calcium carbonate particles. From the
tendency that calcium is widely distributed, it was considered that shear stress
and cohesion could be increased [23–26]. In the Toyoura sand, the shear stress
increased, but the cohesion was decreased. In the Toyoura sand, precipitation
of calcium carbonate on the particle surface was partially precipitated, and
the calcium carbonate played the role of fine grain. After the test, Pca (%) was
measured, and the calcium carbonate precipitation ratio to the sand mass was 1.8
to 2.4% The results are shown in **Table 2**. In the solidification period 1 Week by
the calcium carbonate method, an average of 2% calcium carbonate precipitation
could be confirmed. The slope stabilization has been also carried out by using the
native bacteria and found that the soil strength has been increased significantly
to protect the landslide.

4. Conclusions

It was observed through the X-ray CT and SEM-EDS analysis that the effect of particle size on bio-cementation was great. The amount of calcite generation is more in silica sand than in the Toyoura sand. In addition, it was confirmed that calcium carbonate precipitated was more in the lower part than in the upper part by infiltrating the bacteria and nutrient from the surface. It was evaluated that the precipitation distribution of calcium carbonate inside the specimen could be confirmed by X-ray CT. It was confirmed that from 10 μm to 50 μm calcite of rhombohedral crystal and vaterite of spherical crystal could be confirmed in the crystalline state of calcium carbonate in Toyoura and silica sand, respectively. It was seen the increase of 1.3 to 2.0 times of shear stress after 1 week of biotreated by using *Bacilius Pasturii*. In the cementation period of 1 week an average of 2% calcium carbonate precipitation could be confirmed.

Acknowledgements

The author acknowledged the financial help to conduct this research from the Soil Science Foundation.

Author details

Azizul Moqsud
Yamaguchi University, Ube City, Japan

*Address all correspondence to: azizul@yamaguchi-u.ac.jp

IntechOpen

References

[1] Gowthaman, S., Mitsuyama, S., Nakashima, K., Komatsu, M., Kawasaki, S. Biogeotechnical approach for slope soil stabilization using locally isolated bacteria and inexpensive low-grade chemicals: A feasibility study on Hokkaido expressway soil, Japan. Soils and Foundations, Vol. 59, 2019, pp. 484-499.

[2] DeJong, J.T., Mortensen, B.M., Martinez, B.C., Nelson, D.C., Biomediated soil improvement. Ecol. Eng. Vol. 36, 2010, pp. 197-210.

[3] Moqsud, M.A., Soga, K., Hyodo, M., Nakata, Y. Evaluation of bio-cemented sand for landslide disaster prevention. International symposium on bio geotechnology. September 12-13, Atlanta, USA. 2018.

[4] Bao, R., Li, J., Chen, L. Effect of microbial induced calcite precipitation on surface erosion and scour of granular soils proof of concept. Journal of transportation research board, Vol. 2657, 2017,pp. 10-18.

[5] Stocks-Fisher, S., Galinat, J. K. , Bang, S. S. Microbiological precipitation of CaCO3. Soil Biology and Biochemistry Vol. 31(11),199, pp. 1563-1571.

[6] DeJong, J., Fritzges, M. & Nüsslein, K. Microbial induced cementation to control sand response to undrained shear. Journal of Geotechnical and Geoenvironmental Engineering Vol.132 (11), 2006, pp. 1381-1392.

[7] Gomez, M.G., Anderson, C.M., Graddy, C.M.R., DeJong, J.T., Nelson, D.C., Ginn, T.R. Large scale comparison of bioaugmentation and biostimulation approaches for biocementation of sands. Journal of geotechnical and geoenvironmental engineering, Vol.143, 2017, 4016124.

[8] Jiang, N.J., Soga, K. The applicability of microbially induced calcite

precipitation for internal erosion control in gravel-sand mixtures. Geotechnique, Vol. 67, 2017, pp. 42-55.

[9] Moqsud, M.A. Bioremediation of polluted soil due to tsunami by using recycled waste glass Scientific reports. 11 (14272), 2021

[10] Imran, M.A., Kimura, S., Nakashima, K., Evelpidou, N., Kawasaki, S. Feasibility study of native ureolytic bacteria for biocementation towards coastal erosion protection by MICP method. Applied science, Vol. 9, 2019, pp. 4462-4475.

[11] Jonkers, H.M. Toward bio-based geo and civil engineering for a sustainable society. Procedia Engineering. Vol.171, 2017, pp. 168-175.

[12] Gomez, M.G., Graddy, C., DeJong, J. and Nelson, D. Biogeochemical changes during bio-cementation mediated by stimulated and augmented Ureolytic Microorganisms. Scientific Reports. Vol. 9, 2019, 11517.

[13] Cui, M.J., Zheng, J.J., Zhang, R.J., Lai, H.J., Zhang, J. Influence of cementation level on the strength behavior of bio-cemented sand. Acta Geotech. Vol. 12, 2017, pp. 971-986.

[14] Montoya, B.M., De Jong, J.T. Stress-strain behavior of sands cemented by microbially induced calcite precipitation. J. Geotech. Geoenviron. Eng. Vol. 141.2015

[15] Cheng, L., Cord-Ruwisch, R., Shahin, M.A. Cementation of sand soil by microbially induced calcite precipitation at various degrees of saturation. Can. Geotech. J. Vol. 50,2013, pp. 81-90.

[16] Bareither, C.A., Edil, T.B., Benson, C.H., Mickelson, D.M. Geological and physical factors affecting the friction

angle of compacted sands. J. Geotech. Geoenvironmental Eng. Vol. 134, 2008, pp. 1476– 1489.

[17] Canakci, H., Hamed, M., Celik, F., Sidik, W., Eviz, F. Friction characteristics of organic soil with construction materials. Soils Found. Vol. 56, 2016, pp. 965-972.

[18] DeJong, J.T., Fritzges, M.B., Nu¨sslein, K. Microbially induced cementation to control sand response to undrained shear. J. Geotech. Geoenvironmental Eng. Vol. 132, 2006, pp. 1381-1392.

[19] DeJong, J.T., Mortensen, B.M., Martinez, B.C., Nelson, D.C. Biomediated soil improvement. Ecol. Eng. 36, 2010, pp. 197-210.

[20] Dhami, N.K., Reddy, M.S., Mukherjee, A. Significant indicators for biomineralization in sand of varying grain sizes. Constr. Build. Mater. 104, 2016, pp. 198-207.

[21] Farukh, M.A., Yamada, T.J. Synoptic climatology of winter daily temperature extremes in Sapporo, northern Japan. Int. J. Climatol. 38, 2018, pp. 2230-2238.

[22] Feng, K., Montoya, B.M.,. Quantifying level of microbial-induced cementation for cyclically loaded sand. J. Geotech. Geoenvironmental Eng. Vol. 143, 2017, 06017005.

[23] Feng, K., Montoya, B.M. Influence of confinement and cementation level on the behavior of microbial-induced calcite precipitated sands under monotonic drained loading. J. Geotech. Geoenvironmental Eng. Vol. 142, 2016, 04015057.

[24] Soon, N.W., Lee, L.M., Khun, T.C., Ling, H.S., Factors affecting improvement in engineering properties of residual soil through microbial-induced calcite precipitation. J. Geotech. Geoenvironmental Eng. Vol. 140, 2014, 04014006.

[25] Van Paassen, L.A., Ghose, R., van der Linden, T.J.M., van der Star, W.R. L., van Loosdrecht, M. C. M. Quantifying biomediated ground improvement by ureolysis: large-scale biogrout experiment. J. Geotech. Geoenvironmental Eng. 136, 2010, 1721-1728.

[26] Moqsud, M.A. Slope Soil stabilization through biocementation by native bacteria in chugoku region, Japan. International Journal of Geomate. Vol 21 (81): 36-42.

Section 3

Evaluation of Landslide Processes

Evaluation of Landslide Susceptibility of Şavşat District of Artvin Province (Turkey) Using Machine Learning Techniques

Halil Akinci, Mustafa Zeybek and Sedat Dogan

Abstract

The aim of this study is to produce landslide susceptibility maps of Şavşat district of Artvin Province using machine learning (ML) models and to compare the predictive performances of the models used. Tree-based ensemble learning models, including random forest (RF), gradient boosting machines (GBM), and extreme gradient boosting (XGBoost), were used in the study. A landslide inventory map consisting of 85 landslide polygons was used in the study. The inventory map comprises 32,777 landslide pixels at 30 m resolution. Randomly selected 70% of the landslide pixels were used for training the models and the remaining 30% were used for the validation of the models. In susceptibility analysis, altitude, aspect, curvature, distance to drainage network, distance to faults, distance to roads, land cover, lithology, slope, slope length, and topographic wetness index parameters were used. The validation of the models was conducted using success and prediction rate curves. The validation results showed that the success rates for the GBM, RF, and XGBoost models were 91.6%, 98.4%, and 98.6%, respectively, whereas the prediction rate were 91.4%, 97.9%, and 98.1%, respectively. Therefore, it was concluded that landslide susceptibility map produced with XGBoost model can help decision makers in reducing landslide-associated damages in the study area.

Keywords: landslide susceptibility mapping, machine learning, RF, GBM, XGBoost, Şavşat

1. Introduction

Natural disasters cause displacement of people, injuries, loss of life, and damage to infrastructure facilities and cultural heritage, which can directly give rise to extreme economic losses. According to the data from Emergencies Database (EM-DAT), managed by the Center for Research on the Epidemiology of Disasters (CRED), 11,755 people died worldwide due to 396 natural disasters that occurred in 2019; 94.9 million people were affected by these disasters and an economic loss of 103 billion dollars was suffered [1]. On the contrary, according to the report prepared by the AON company, which provides insurance and reinsurance brokerage and risk management consultancy services, the damage caused by natural disasters in 2020 is estimated to be 268 billion dollars [2]. In the AON report prepared in 2020, the value of total economic losses caused by natural disasters in the 2010–2019 period was calculated as 2.98 trillion dollars. In the same report, the economic losses in question were reported to be 1.1 trillion dollars higher than that in the 2000–2009 period [3].

Landslide is generally defined as the downward movement and displacement of the material forming a slope with the effect of gravity [4]. Rabby and Li [5] stated in their study that landslides are a very common phenomenon and account for 9% of disasters in the world. Landslides, especially those caused by rainfall, are the most damaging natural disasters in mountainous and rugged regions, resulting in loss of life, damage to property, and economic loss [6]. Landslide susceptibility maps are one of the important data needed to identify landslide-hazardous areas and to reduce losses due to landslides [7, 8]. Many different approaches and models have been implemented in the production of landslide susceptibility maps. Merghadi et al. [9] and Tang et al. [10] classified the modeling approaches into four categories: the heuristic, physically based, statistical, and machine learning (ML) models. Heuristic and physically based models (also known as deterministic models) have their own characteristics and disadvantages. Heuristic models are highly subjective and rely on experts' opinions and experience on assigning weightage to landslide-conditioning factors [11–14]. In this approach, differences in expert opinions or insufficient information about the study area may cause inconsistent results [15]. Physically based or deterministic models use laws of mechanics to analyze slope stability. The advantages of these models are that they do not require long-term landslide inventory data and are more useful in areas where landslide inventories are missing [15]. However, deterministic models are suitable for small areas where landslide types are simple and ground conditions are fairly uniform [14], but they require detailed geotechnical and hydrogeological data on these areas [13]. To overcome the disadvantages of the above two approaches and to produce reliable landslide susceptibility maps, statistics-based models have been developed [14]. Statistics-based models evaluate the correlation between past landslides and the conditioning factors that had an impact on their occurrence [16] and they need landslide inventory data for this [17].

In recent years, machine learning (ML) techniques such as support vector machine [18, 19], decision tree [20, 21], generalized linear model [22, 23], logistic model tree [13, 16], artificial neural networks [6, 24, 25], and Naïve Bayes [26–28] have been widely applied for landslide susceptibility mapping (LSM). Sahin [29] and Merghadi et al. [9] stated that tree-based ensemble algorithms provide better prediction performance for LSM compared to any single model. In addition, Sahin [30] stated that ensemble learning techniques, such as random forest (RF), gradient boosting machine (GBM), and extreme gradient boosting (XGBoost), are efficient and robust for creating landslide susceptibility maps and that these algorithms would be preferred more frequently in the future for their robustness.

The most common natural disasters in Turkey are landslides and floods. Artvin is one of the provinces in Turkey that experiences the most frequent natural disasters. Landslides occur almost every year in the province of Artvin, especially due to meteorological conditions (extreme rainfall) and anthropogenic activities, such as agricultural activities, excessive irrigation, and road excavations. Şavşat is one of the districts of Artvin where landslides are most common. Şavşat, a Cittaslow city, stands out with its historical and natural beauties and has a high tourism potential. For this reason, it is very important to evaluate the landslide susceptibility to reduce the landslide-associated damages in the district. The aim of this study is to produce landslide susceptibility maps of Şavşat district of Artvin Province using RF, GBM, and XGBoost ML models and to evaluate the performances of the models. Eleven factors commonly used in LSM studies were used in the study. The produced landslide susceptibility maps were validated using the validation dataset.

2. Study area and data used

Şavşat, like other districts of Artvin, is a district with a rugged terrain. Şavşat, spreading on a 1272.27 km² land, is located between 41°05′11″ and 41°30′56″ north latitudes and 42°04′30″ and 42°35′47″ east longitudes (**Figure 1**). In the study area, the altitude varies between 590 and 3005 m with the average altitude being 1789.14 m. The average slope of the study area is 21.17°, whereas the maximum slope is 72.53°. The slope is over 20° in ~55% of the study area.

According to the data from the Turkish Statistical Institute (TURKSTAT), the total population of Şavşat district in 2020 is 17,024. Of this population, 6,123 live in the town and 10,901 live in villages [31]. There is a transitional climate between the Black Sea climate and the continental climate in the district. While semi-humid climatic conditions are observed in the low valley floors, cold humid climatic conditions are observed in the higher elevations. In addition, winters are very long in places with high altitudes. According to the data (November 2012–March 2021) from the General Directorate of Meteorology, sum of monthly average rainfall in the study area is 715.60 mm. The monthly average rainfall is minimum in February with 27.8 mm and maximum in May with 111.03 mm. In the study area, the monthly average temperature was maximum at 32.8°C in August and minimum at −7.4°C in December [32].

Şavşat is located in the eastern part of the Eastern Pontides and the southern part of Transcaucasia. In the study area, intrusive, volcanic, and volcano-sedimentary facies have developed due to the magmatic activities that took place in the Dogger, Late Cretaceous, and Eocene ages. In the north and northwest part of the region, units representing the same stratigraphic unity surfaces in a range extending from the Liassic to the Early-Middle Eocene. In the southern part, units representing two separate stratigraphic units are surfaced. The sequence in the west of the southern section is characterized by units of Early-Middle Jurassic and Late

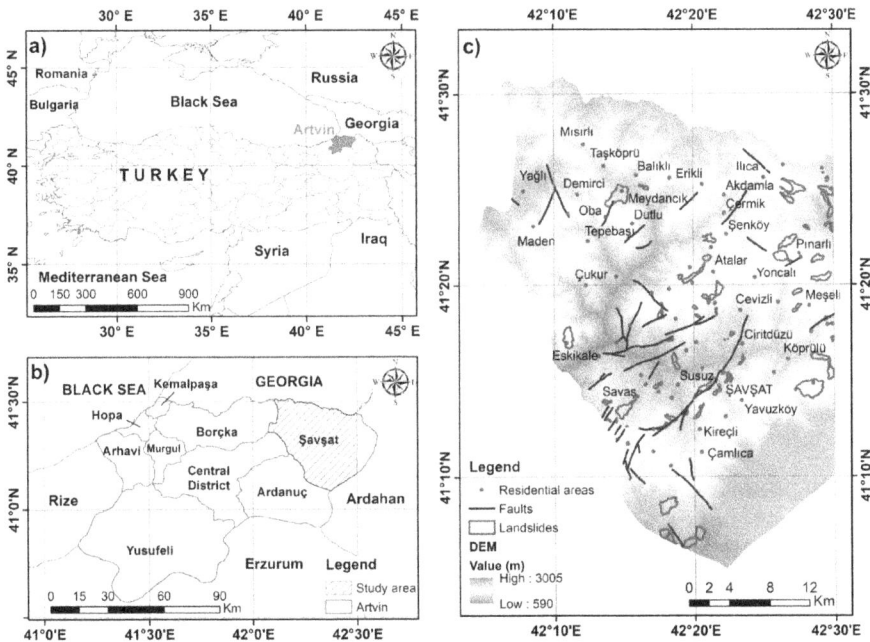

Figure 1.
Study area.

Cretaceous age, and the sequence in the east of the southern section is characterized by units of Late Cretaceous and Middle Eocene age. Tertiary units surfacing in the eastern and southeastern parts of the region are considered as common units [33]. According to the earthquake zone map of Turkey, Şavşat district is located in the third degree earthquake zone. However, the most common natural disaster in the district is landslide [34]. The landslides occurring in the study area are mostly of complex type. Landslides are observed in larger areas with respect to Quaternary alluvium and slope debris [33].

2.1 Landslide inventory map

To reliably predict future landslides, reliable landslide inventory maps containing information about past landslides are needed [16]. As stated by Parise [35], landslide inventory maps represent the spatial distribution of landslides and provide information about the location, typology, and activity status of landslides. In this study, the landslide inventory map produced by Artvin Provincial Directorate of Disaster and Emergency was used. The landslide inventory map contains 85 landslide polygons. The area of the smallest landslide polygon in the study area is 0.01 ha (99.34 m^2), and the area of the largest landslide polygon is 325.97 ha. The average area of the landslide polygons is 34.75 ha. Landslides cover ~3% of the study area. The lengths of the landslides in the region vary between 13 and 3100 m and their widths vary between 10 and 2780 m. According to their activities, 28 of these landslides are active, 32 are stalled, and 25 are inactive landslides. According to Varnes [4] classification of mass movements, 6 of the landslides were classified as slide, 2 as lateral spread, 20 as flow, and the remaining 57 as complex.

2.2 Landslide-conditioning factors

Evaluation of landslide susceptibility in a region depends on determining the factors that are effective in the formation of landslides in that region and on collecting spatial data related to these factors [36]. Yi et al. [8] stated that there is no widely accepted procedure for the selection of factors used in LSM. Yanar et al. [37], on the contrary, stated that the main limitation in determining the factors to be used to create landslide susceptibility maps is the availability of data. In this study, 11 factors including altitude, aspect, curvature, distance to drainage network, distance to faults, distance to roads, land cover (CORINE 2018), lithology, slope, slope length, and topographic wetness index (TWI) were used based on the availability of data, geo-environmental conditions of the study area, and literature survey. Spatial data on these factors are collected from different sources (**Table 1**). Landslide-conditioning factor maps were generated using ESRI ArcGIS 10.5 and SAGA GIS 7.9.0 software and were converted into raster format with 30 m spatial resolution.

2.2.1 Altitude

Altitude is associated with various geomorphological and meteorological factors such as weathering, weather conditions, wind effect, and precipitation, which are effective in the formation of landslides [6]. For this reason, it has been used in almost all LSM studies. The digital elevation model (DEM) of the study area was created using 10-m-interval contours on the topographic maps and it was converted to raster format with 30-m spatial resolution. The altitude map of the study area was generated from this DEM. The altitude in the study area varies between 590 and 3005 m. DEM was reclassified into 10 classes at 240 m intervals (**Figure 2a**).

Original data	Factors	Data type	Scale	Data provider
Landslide inventory	Landslide locations	Polygon	1/25,000	Artvin Provincial Directorate of Disaster and Emergency
Geological map	Lithology	Polygon	1/100,000	General Directorate of Mining Research and Exploration (GDMRE)
	Distance to fault lines	Polyline	1/100,000	
Topographical map	Altitude	GRID	1/25,000	General Directorate of Mapping
	Slope	GRID	1/25,000	
	Slope length	GRID	1/25,000	
	Aspect	GRID	1/25,000	
	Curvature	GRID	1/25,000	
	TWI	GRID	1/25,000	
	Distance to drainage network	GRID	1/25,000	
Road network	Distance to roads	Polyline	1/25,000	Basarsoft Information Technologies Inc.
CORINE 2018	Land cover	Polygon	1/100,000	European Union Copernicus Land Monitoring Service

Table 1.
Data and data sources.

2.2.2 Aspect

Aspect has an important role in landslide formation as it affects factors such as exposure to sunlight and the intensity of solar radiation, wind, rainfall and, soil moisture [38, 39]. For this reason, aspect is widely used in LSM studies [6, 26, 36, 40]. The aspect map used in this study was produced from DEM and divided into nine classes (flat, north, northeast, east, southeast, south, southwest, west, and northwest) (**Figure 2b**).

2.2.3 Curvature

Curvature, which is widely used in geomorphometric analysis, is one of the basic terrain parameters and reflects the shape of the land surface [23, 41]. In curvature map, positive curvature values indicate that the surface is convex, negative curvature indicates that the surface is concave, and zero indicates that the surface is flat [42]. In this study, curvature map was derived from DEM using ArcGIS 10.5 software and divided into three subclasses, i.e., concave, flat, and convex (**Figure 2c**).

2.2.4 Distance to drainage network

The distance to the drainage networks is one of the important conditioning factors used in landslide susceptibility studies, since the pore water pressure that causes the formation of landslides increases in areas close to the drainage networks [23]. Drainage networks in the study area were generated from DEM using functions in ArcHydro toolbox in ArcGIS 10.5 software. The distance to the drainage networks was calculated using the Euclidean distance tool in ArcGIS 10.5.

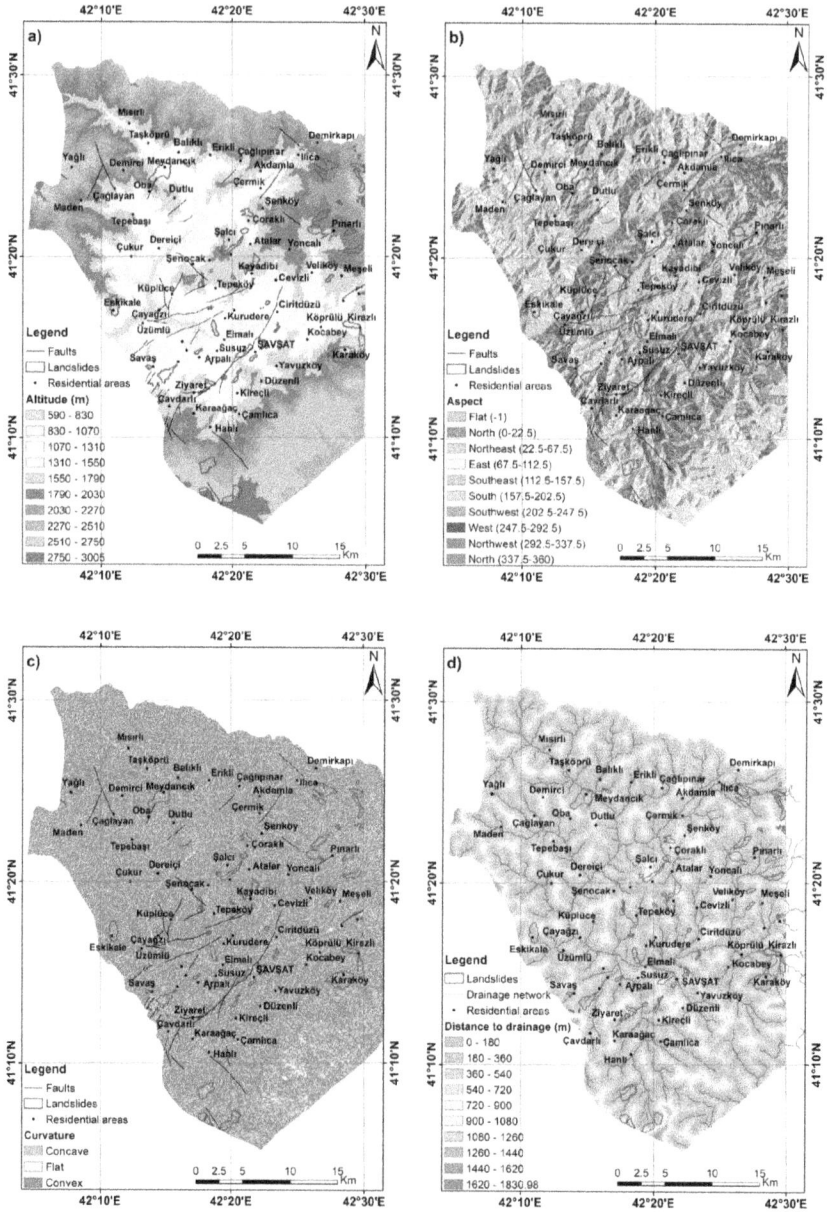

Figure 2.
The landslide conditioning factor maps: a) altitude b) aspect c) curvature d) distance to drainage network.

The maximum distance to the drainage networks in the study area has been calculated as 1830.98 m. The distance to the drainage networks is reclassified into 10 subclasses with equal intervals of 180 m (**Figure 2d**).

2.2.5 Distance to faults

Areas close to faults are highly susceptible to landslides as the strength decreases due to tectonic fractures [28]. Ba et al. [43] stated that landslides tend to occur around faults due to fractures in the rock mass. For this reason, the distance to the faults is taken into account in the landslide susceptibility analysis [14, 40, 44]. In this study, the distance to the faults was obtained using the Euclidean distance tool of ArcGIS 10.5 software. The maximum distance to the faults in the study area has been calculated as 13,016.61 m. The distance to the faults was classified into 10 subclasses with 1200 m intervals and used in the landslide susceptibility analysis (**Figure 3a**).

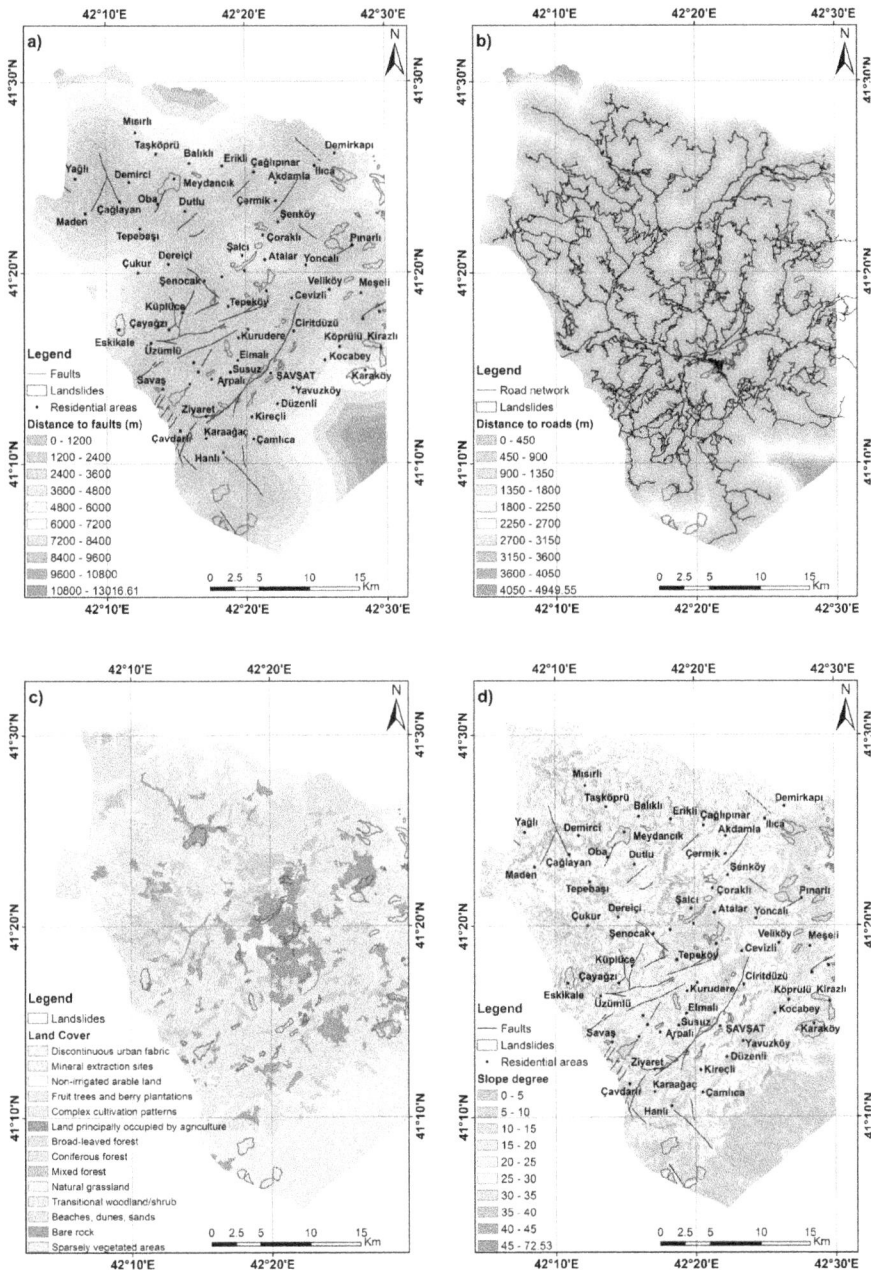

Figure 3.
The landslide conditioning factor maps: a) distance to faults b) distance to roads c) land cover d) slope.

2.2.6 Distance to roads

Road construction, which is considered to be one of the most important anthropogenic factors, destabilizes the slopes, so the probability of landslides along a road increases [43]. Roads built on slopes in areas with rough topography cause loss of toe support, change in topography, increase in tension behind the slope, and development of tension cracks [45, 46]. For this reason, distance to the road has been considered as one of the important conditioning factors in many studies [14, 17, 47]. The road network in the study area was supplied in digital format from Başarsoft Information Technologies Inc., which collects road data for the production of navigation maps in Turkey. Distance to roads was calculated using the Euclidean distance tool in ArcGIS 10.5 and reclassified into 10 subclasses at 450 m intervals (**Figure 3b**).

2.2.7 Land cover

Land cover maps, in general, represent what physical classes or materials (e.g., forest, pasture, field, lake, and wetland) the Earth's surface is spatially covered with. Land use or land cover maps are usually used in LSM studies for taking into consideration the effects of anthropogenic activities on rugged slopes on landslide formation [5]. In this study, CORINE 2018 land cover (CLC 2018) data provided by Copernicus Land Monitoring Service, one of the European Union's Earth Observation Programme services, were used. According to this dataset, the study area includes 14 different land cover classes (**Figure 3c**).

2.2.8 Slope

The slope angle, one of the most important factors governing the stability of slopes, is closely related to the shear forces acting on the slopes. As the angle of

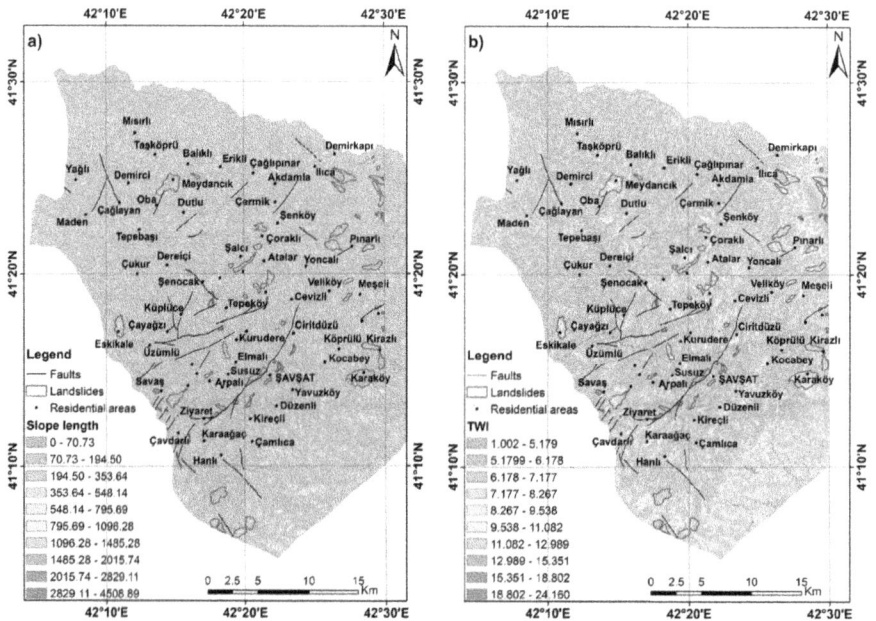

Figure 4.
The landslide conditioning factor maps: a) slope length b) TWI.

inclination increases, the shear stress in the materials forming the slope generally increases [48]. For this reason, slope angle has been used in all LSM studies, as is the case for the lithology parameter [18, 40, 49–51]. The slope in the study area varies between 0° and 72.53°. In this study, the slope was divided into 10 classes with 5° spacing, and a slope map of the study area was produced (**Figure 3d**).

2.2.9 Slope length

Slope length is one of the important topographic factors that affect the formation of landslides [6]. Kavzoglu et al. [18] defines the slope length as "the distance along a slope subject to uninterrupted over land flow." Slope length affects hydrological processes and soil loss, especially in mountainous areas [23]. This factor is closely related to the formation of landslides, because the potential for the materials forming the slopes to be carried downhill also increases with the increase of the slope length [52]. In this study, slope length was produced from DEM using SAGA GIS software and it was reclassified into 10 classes using the natural break classification method (**Figure 4a**).

2.2.10 Topographic wetness index (TWI)

TWI is an index generally used to characterize the spatial distribution of soil moisture [53] and is considered as an important factor contributing to the occurrence of landslides. Yanar et al. [37] stated that TWI indicates the locations and size of the water-saturated regions. For this reason, TWI has been used in many landslide susceptibility studies [26, 54, 55]. The following equation is used to calculate TWI:

$$TWI = \ln\left(\frac{A_s}{tan\beta}\right) \tag{1}$$

In the Eq. (1), As is the specific basin area and β is the slope in degrees. TWI index in the study area, varying between 1.002 and 24.160, was produced using SAGA GIS software. TWI index values were divided into 10 subclasses using the natural break classification method and used in sensitivity analysis (**Figure 4b**).

2.2.11 Lithology

Kavzoglu et al. [18] stated that lithology is one of the main factors that have a direct impact on the formation of landslides, as lithological and structural variations lead to changes in the strength and permeability of rocks and soils. For this reason, lithology has been one of the most important conditioning factors used in all landslide susceptibility evaluation studies. In this study, 1/100,000 scaled digital geological map obtained from General Directorate of Mineral Research and Exploration (GDMRE) was used to produce the lithological map of the study area. The geological map of the study area includes 16 lithological units (**Figure 5**).

3. Methodology

3.1 Random forest

First proposed by Breiman [56], RF is an ensemble learning method that creates multiple decision trees from the training dataset and combines the results of

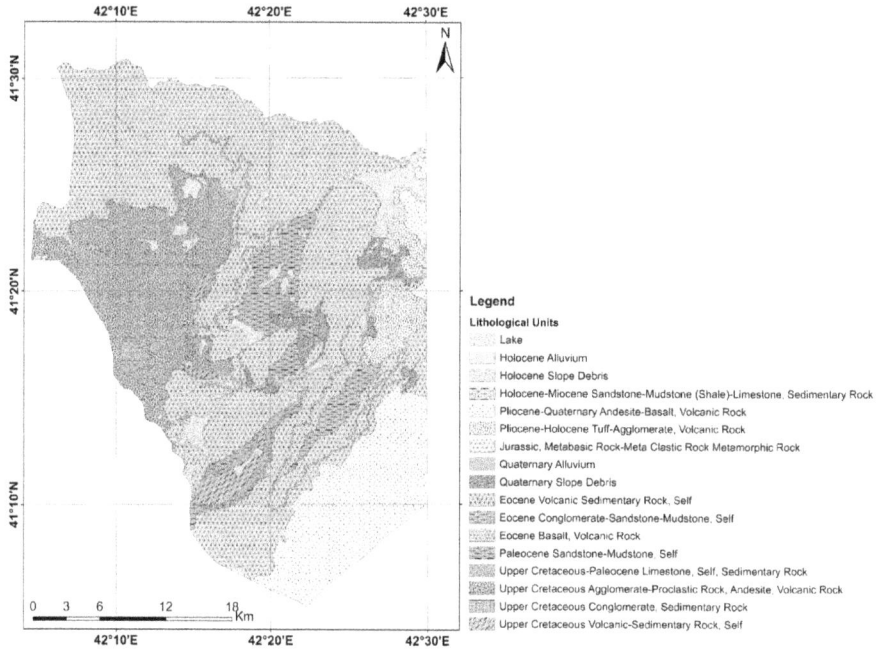

Figure 5.
Lithological map of the study area.

the decision trees to improve the predictive ability of the model [57]. According to Arabameri et al. [44] and Merghadi et al. [9], one of the most important advantages of RF is that it avoids the risk of overfitting, which is a common problem in other decision tree models. In the study conducted by Sahin [29], it is stated that requiring less hyperparameter tuning, compared to gradient boosting algorithms, was RF's main advantage. To create a classification model in RF, two parameters must be defined: *ntree* parameter, which refers to the number of decision trees generated by RF, and *mtry* parameter, which refers to the number of factors or variables used in each node of the decision tree. In this study, "*rf*" method of the "*caret*" package [58] was used in R 3.6.3 to apply the RF model. In the study, the *ntree* parameter was set to 100 and the *mtry* parameter to 8, and a 10-fold cross validation approach was used to reduce the variability of the model results.

3.2 Gradient boosting machine (GBM)

GBM [59] is a ML technique that combines multiple different models through boosting and regression trees to increase prediction precision [60]. The main feature of GBM is that it combines multiple weak learners to improve their performances. GBM, an ensemble learning method, combines multiple decision trees to create a more powerful model that can be used for classification or regression. In GBM, unlike RF, each tree tries to correct the error of the previous tree [61]. For this purpose, the residual errors calculated as a result of the prediction of the previous tree are minimized and the next tree is obtained, and these processes continue until the prediction results are stable or until the maximum number of trees is reached. In practice, the number of trees is chosen to be 100 or greater. There are four parameters that must be set by the user during the execution of the GBM, namely number of trees (*n.trees*), *shrinkage*, number of levels of trees (*interaction.depth*), and the minimum number of observations in trees' terminal nodes (*n.minobsinnode*). For low-variance and accurate predictions, the learning rate is chosen so that it converges to the optimum value with small steps in the right direction. The number of levels of trees is chosen between 8 and 32. In this study, GBM was performed using the *"gbm"* method in R 3.6.3.

3.3 Extreme gradient boosting (XGBoost)

XGBoost, developed by Chen and Guestrin [62], is based on the gradient boosting approach. XGBoost is based on the efficient and effective implementation of the gradient boosting algorithm. For this purpose, it interprets the approximate greedy algorithm with the Newton–Rapson method. XGBoost uses several classification and regression trees and integrates them using gradient boosting [63]. It produces fast and accurate solutions with univocal regression trees, weighed quantile approach, and sparsity aware split finding. It is trained very quickly, and since it is suitable for parallel learning technique, XGBoost increases the overall accuracy (performance) of the model by avoiding the overfitting problem during the training process [64]. XGBoost uses two additional techniques called shrinkage and column (feature) subsampling to avoid overfitting [62]. Wang et al. [61] noted that the computational speed and accuracy of XGBoost has been significantly improved compared to GBM. In this study, the XGBoost model is implemented in R 3.6.3 using the *"xgbTree"* method of the *"caret"* package.

3.4 Preparation of training and validation dataset

"Landslide (or positive)" and "non-landslide (or negative)" samples are needed in the study area during the training and validation of the models used to create landslide susceptibility maps. The ratio of 70:30 has been commonly used in the literature to produce training and validation datasets [6, 8, 65, 66]. In particular, 70% of the landslide inventory data is used for training the models and the remaining 30% is used for the validation of the models. Huang and Zhao [67], on the contrary, stressed that the number of positive and negative samples in the training and validation datasets should be equal, i.e., having a ratio of 1:1. For this reason, as many negative samples as the number of positive samples are selected in the study area. In this study, 85 landslide polygons on the inventory map were converted to 30 m × 30 m resolution raster format and 32,777 landslide pixels were obtained. A value of "1" was assigned to positive or landslide pixels in the study area. Then, 32,777 non-landslide pixels were randomly selected in the study area in the R program and the value of "0" was assigned to these pixels. Randomly selected 70% of the landslide and non-landslide pixels (45,888 pixels in total) were used for training the models and the remaining 30% (19,666 pixels) were used for the validation of the models.

3.5 Multicollinearity analysis for landslide-conditioning factors

One of the important steps of LSM is to control the multicollinearity between landslide-conditioning factors [8]. Multicollinearity is an important analysis used to determine the conditional independence between the factors during the selection of the conditioning factors to be used in susceptibility models, and thus, to prevent the models from producing erroneous predictions [9, 68]. Commonly used indicators for multicollinearity analysis are tolerance (TOL) and variance inflation factor (VIF). A TOL value less than 0.1 or a VIF value greater than 10 indicates multicollinearity [8, 16, 44]. TOL and VIF values calculated using the training dataset for this study are shown in **Table 2**. The results show that there is no multicollinearity

Landslide conditioning factors	Statistics	
	TOL	VIF
Altitude	0.4713	2.1217
Aspect	0.9770	1.0235
Curvature	0.7879	1.2692
Distance to drainage network	0.7916	1.2633
Distance to faults	0.7786	1.2844
Distance to roads	0.5552	1.8011
Land cover	0.7206	1.3877
Lithology	0.8763	1.1412
Slope	0.5373	1.8610
Slope length	0.7345	1.3615
Topographic Wetness Index	0.4595	2.1761

Table 2.
Multicollinearity analysis of landslide-conditioning factors.

among the landslide-conditioning factors used in the study. Therefore, all selected factors were used to produce landslide susceptibility map of the study area.

4. Results and discussion

4.1 Landslide susceptibility mapping

In this study, RF, GBM, and XGBoost models were successfully applied and landslide susceptibility index (LSI) maps were produced via R 3.6.3 using the training data set for each model. Then, landslide susceptibility maps were obtained by reclassifying the LSI maps into five classes: very low, low, medium, high, and very high, using the natural breaks (Jenks) classification method in ArcGIS 10.5 software (**Figure 6**).

The spatial distributions (in percentages) of the susceptibility classes for each model are given in **Figure 7**. It has been determined that the study area is highly or very highly susceptible to landslides by 27.27%, 11.13%, and 16.89% according to the GBM, RF, and XGBoost models, respectively (**Figure 7**).

The significance degrees of the landslide-conditioning factors used in the study are presented in **Figure 8**. It has been observed in all models that the lithology is the most important parameter. After lithology, the most important or most effective parameters in the study area were determined to be altitude, distance to faults, slope, and land cover parameters. Slope length and curvature were the least significant parameters in all models (**Figure 8**). The findings related to the parameters found to be effective in terms of landslide are explained in the following sections.

When **Table 3** is examined, ~76% of the landslides in the study area can be seen to have occurred at altitudes between 1070 and 2030 m. In respect of altitude, 1070–1310, 1310–1550, 1550–1790, and 1790–2030 m altitude classes were found to be susceptible to landslides (**Table 3**). The main reason why these altitude classes are susceptible to landslides is that more than 90% of the village settlements in the study area are located between these altitudes. Uncontrolled excavations and uncontrolled agricultural activities in villages are the most important factors that trigger landslides. In the study by Erener et al. [34], conducted in Şavşat district and

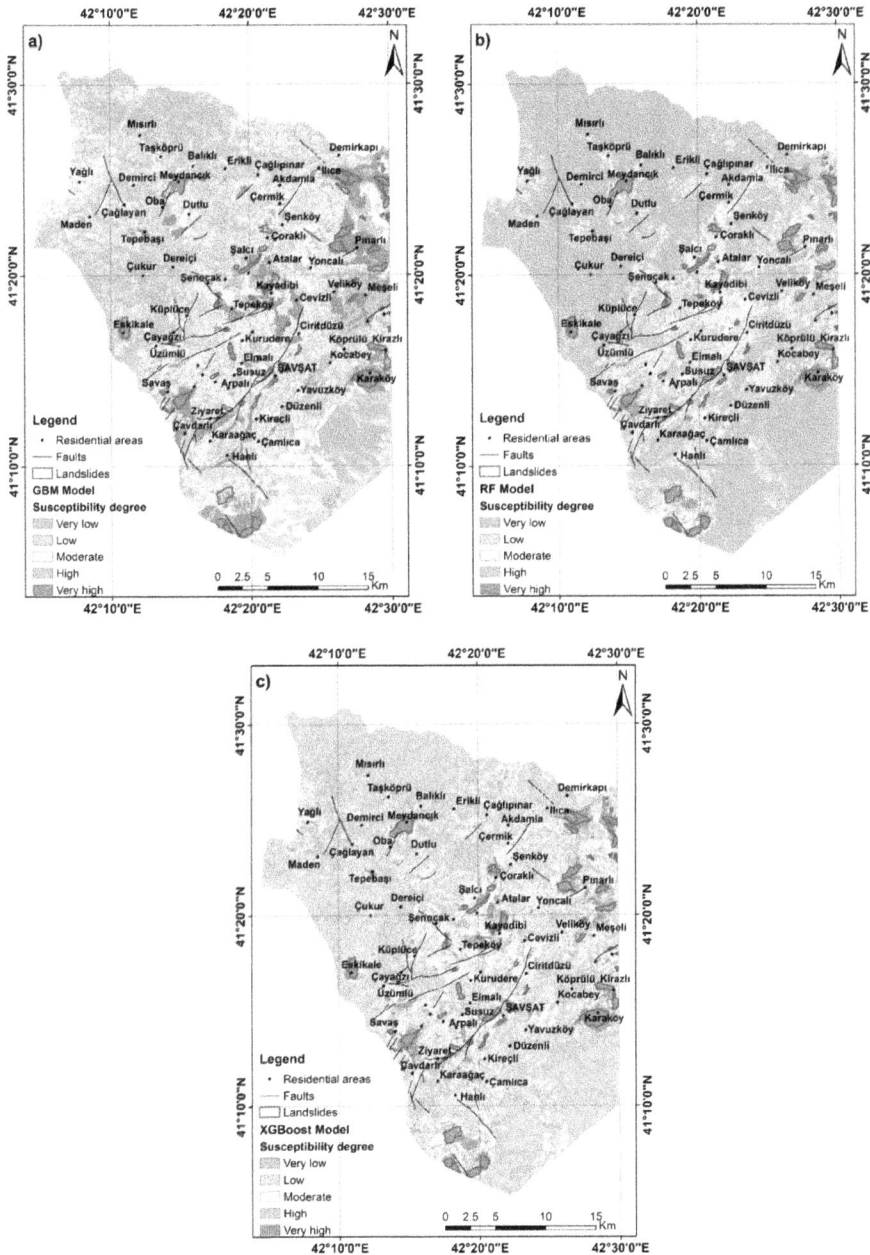

Figure 6.
Landslide susceptibility maps produced using a) GBM b) RF c) XGBoost.

covering a more limited (small) region compared to this study, the altitude class between 1500 and 2000 m was found to be susceptible to landslides.

When the study area is examined in terms of slope, it is seen that 0°–5°, 5°–10°, 10°–15°, and 15°–20° slope classes are more susceptible to landslides (**Table 3**). In these slope classes, 82.31% of the landslides occurred in the study area. The fact that complex mass movements (creeping and spreading) in the study area are generally seen in areas with low slope degrees (approximately in the range of 7°–12°) have provided these results in terms of slope.

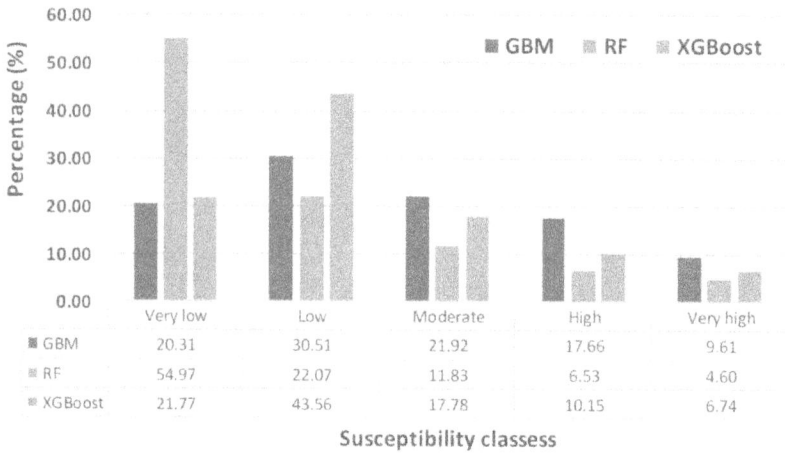

Figure 7.
Percentage distributions of susceptibility classes.

	Very low	Low	Moderate	High	Very high
GBM	20.31	30.51	21.92	17.66	9.61
RF	54.97	22.07	11.83	6.53	4.60
XGBoost	21.77	43.56	17.78	10.15	6.74

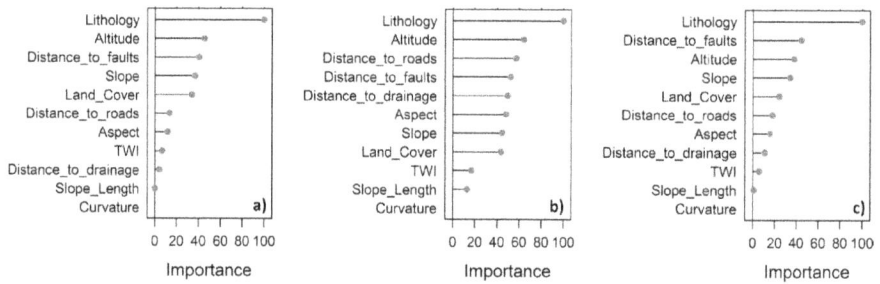

Figure 8.
Importance of landslide-conditioning factors for a) GBM b) RF c) XGBoost.

When **Table 3** is examined, it is seen that ~55% of the landslides in the study area occur on slopes with north, northeast, and northwest aspects. When the frequency ratios in **Table 3** are examined, it is clearly seen that the slopes with these aspects have the highest frequency ratio value, and therefore, they are more susceptible to landslides. In the study conducted by Akıncı and Zeybek [69], in the Ardanuç district, which is adjacent to the Şavşat district and has similar topographical and geomorphological characteristics with the study area, the slopes with north, northwest, and northeast aspects were determined to be more susceptible to landslides.

Within the first 3600 m margin of the faults, 74% of the landslides occurred in the study area (**Table 3**). In the study area, the landslide susceptibility tends to decrease with distance from the faults. Although the region most susceptible to landslides in terms of distance to faults is 4800–6000 m, it is seen that distance classes of 0–1200, 1200–2400, and 2400–3600 m are also susceptible to landslides (**Table 3**). Althuwaynee et al. [70] stated that the probability of landslide decreases as the distance to the faults increases. Also in the LSM study conducted by Akinci et al. [40] in the area covering Arhavi, Hopa, and Kemalpaşa districts of Artvin Province, the areas within the first 2000 m distance to the faults were determined to be more susceptible to landslides.

Considering the CORINE 2018 land cover data, it was determined that ~56% of the landslides in the study area occurred in agricultural areas (**Table 3**).

Factor	Subclasses	Pixels in domain	Pixels with landslide	Percentage of landslides (%)	Percentage of domain (%)	FR
Altitude (m)	590–830	16659	0	0.00	1.34	0.0000
	830–1070	65715	928	2.83	5.28	0.5363
	1070–1310	129711	3488	10.64	10.42	1.0212
	1310–1550	202140	5893	17.98	16.24	1.1071
	1550–1790	242416	7455	22.74	19.48	1.1679
	1790–2030	188846	8231	25.11	15.17	1.6552
	2030–2270	140243	1174	3.58	11.27	0.3179
	2270–2510	120268	2402	7.33	9.66	0.7585
	2510–2750	121891	1687	5.15	9.79	0.5256
	2750–3005	16840	1519	4.63	1.35	3.4255
Slope (degree)	0–5	86650	3307	10.09	6.96	1.4493
	5–10	156262	10914	33.30	12.55	2.6524
	10–15	160085	8205	25.03	12.86	1.9464
	15–20	160639	4554	13.89	12.91	1.0766
	20–25	174982	2652	8.09	14.06	0.5756
	25–30	194812	1656	5.05	15.65	0.3228
	30–35	176843	1069	3.26	14.21	0.2296
	35–40	96265	302	0.92	7.73	0.1191
	40–45	28481	98	0.30	2.29	0.1307
	45–72.53	9710	20	0.06	0.78	0.0782

Factor	Subclasses	Pixels in domain	Pixels with landslide	Percentage of landslides (%)	Percentage of domain (%)	FR
Aspect	Flat	4416	49	0.15	0.35	0.4214
	North	148077	6845	20.88	11.90	1.7555
	Northeast	151999	5873	17.92	12.21	1.4673
	East	148757	3387	10.33	11.95	0.8647
	Southeast	161166	2816	8.59	12.95	0.6635
	South	162974	2667	8.14	13.09	0.6215
	Southwest	161749	2269	6.92	12.99	0.5327
	West	155008	3680	11.23	12.45	0.9016
	Northwest	150583	5191	15.84	12.10	1.3091
CORINE 2018	112	861	45	0.14	0.07	1.9848
	131	333	0	0.00	0.03	0.0000
	211	927	275	0.84	0.07	11.2657
	222	408	0	0.00	0.03	0.0000
	242	133175	7439	22.70	10.70	2.1213
	243	130192	10747	32.79	10.46	3.1348
	311	46639	241	0.74	3.75	0.1962
	312	340177	3839	11.71	27.33	0.4286
	313	96234	82	0.25	7.73	0.0324
	321	278125	6054	18.47	22.34	0.8266
	324	149283	2678	8.17	11.99	0.6812
	331	1380	0	0.00	0.11	0.0000
	332	4746	181	0.55	0.38	1.4483
	333	62249	1196	3.65	5.00	0.7296

Factor	Subclasses	Pixels in domain	Pixels with landslide	Percentage of landslides (%)	Percentage of domain (%)	FR
Distance to faults (m)	0–1200	353573	10042	30.64	28.41	1.0786
	1200–2400	311578	8822	26.92	25.03	1.0752
	2400–3600	198557	5463	16.67	15.95	1.0448
	3600–4800	132651	2932	8.95	10.66	0.8394
	4800–6000	91279	3955	12.07	7.33	1.6454
	6000–7200	62754	1563	4.77	5.04	0.9459
	7200–8400	43119	0	0.00	3.46	0.0000
	8400–9600	25977	0	0.00	2.09	0.0000
	9600–10800	14205	0	0.00	1.14	0.0000
	10800–13016.61	11036	0	0.00	0.89	0.0000

Factor	Subclasses	Pixels in domain	Pixels with landslide	Percentage of landslides (%)	Percentage of domain (%)	FR
Lithology	Lake	915	0	0.00	0.07	0.0000
	e-10-s	625072	7853	23.96	50.22	0.4771
	e-18-s	64071	1795	5.48	5.15	1.0639
	e-V2	15193	6	0.02	1.22	0.0150
	Jbmclm	524	0	0.00	0.04	0.0000
	k2-10-s	19698	77	0.23	1.58	0.1484
	k2-2-k	256	0	0.00	0.02	0.0000
	k2-pn-8-s	29136	1800	5.49	2.34	2.3461
	k2-V16-V15-V13	189788	942	2.87	15.25	0.1885
	plQ-V13-V2	163657	6297	19.21	13.15	1.4612
	plQ2-V17-V16	41251	1883	5.74	3.31	1.7335
	pn-19-s	10873	383	1.17	0.87	1.3377
	Q-21-k	6144	16	0.05	0.49	0.0989
	Q-23-k	23921	4063	12.40	1.92	6.4502
	Q2-21-k	434	0	0.00	0.03	0.0000
	Q2-23-k	50601	7359	22.45	4.07	5.5229
	Q2m-20-ks	3195	303	0.92	0.26	3.6014

Table 3.
Spatial relationship between landslide-conditioning factors and landslides.

Non-irrigated arable lands (CORINE land cover code 211), agricultural areas within natural vegetation (243), mixed agricultural areas (242), discontinuous urban structure (112), and bare rocks (332) were determined as landslide sensitive areas. The scattered settlements in the villages cause uncontrolled excavations, which in turn triggers landslides. In the landslide susceptibility study conducted by Erener et al. [34] in Şavşat district, it was reported that landslide activity increased in areas where the original vegetation was removed or changed. In the same study, it was determined that farming areas, irrigated or dry, were more susceptible to landslides. Researchers attributed this to the deforestation in agricultural areas.

4.2 Validation and comparison of landslide susceptibility models

Thi Ngo et al. [7] stated that it is important to identify landslide-prone areas with high accuracy and to use an appropriate metric for the performance evaluation to produce a reliable landslide susceptibility map. The performances of the models used in the production of landslide susceptibility maps are mostly evaluated using the receiver-operating characteristics (ROC) curve [28, 38, 45, 60, 71–73]. Therefore, in this study, the receiver-operating characteristic-area under the curve (ROC-AUC) approach was applied to evaluate and measure the performances of ML models. The ROC curve is a graph showing the true positive rate (TPR or sensitivity) on the vertical axis and the false positive rate (FPR or 1-specificity) on the horizontal axis. In the ROC curve, the most important indicator used to evaluate the accuracy or performance of the susceptibility model is the AUC. AUC takes values between 0.5 and 1 [71]. An AUC value close to 1.0 indicates high performance of the model and close to 0.5 indicates low performance of the model. On the contrary, Chen et al. [74] and Wang et al. [17] stated that the AUC value can be classified in five classes: poor (0.5–0.6), moderate (0.6–0.7), good (0.7–0.8), very good (0.8–0.9), and excellent (0.9–1.0).

In the study, success rate and prediction rate curves were created using training and validation data sets, respectively. The success rate curve is used to understand how well the models used to produce landslide susceptibility maps to classify existing landslide areas [74]. In this study, the AUC values of the success rate curves for the GBM, RF, and XGBoost models were calculated as 91.6%, 98.4%, and 98.6%, respectively (**Figure 9a**). Since the success rate curve is produced using the training

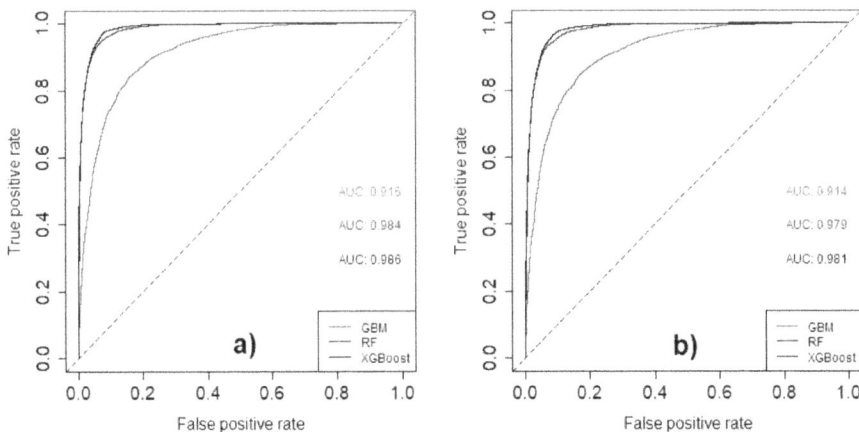

Figure 9.
a) Success rate b) prediction rate curves for ML models.

data set, it is not an appropriate indicator to evaluate the predictive capabilities of the models [21, 42]. The prediction rate curve should be used to evaluate the prediction capabilities of the models [75]. The prediction rate curve shows how well the models predict unknown or probable future landslides [5]. The AUC values of the prediction rate curves produced for the GBM, RF, and XGBoost models were calculated as 91.4%, 97.9% and 98.1%, respectively (**Figure 9b**). AUC value being close to 1.0 in three models show, according to the classification made by Chen et al. [74] and Wang et al. [17], that their performances, i.e., their prediction capacities, are excellent.

5. Conclusions

In this study, RF, GBM, and XGBoost algorithms were used for landslide susceptibility mapping of Şavşat district of Artvin Province. The performances of these models were evaluated using success rate and prediction rate curves. According to the AUC values, the models used in the study showed excellent performance. However, the XGBoost model outperformed the other two models in landslide susceptibility mapping of the study area. Therefore, it was concluded that the susceptibility map produced by the XGBoost model can help decision makers and planners in reducing the risks caused by landslides in the region and in land use planning. In this study, 11 factors—altitude, aspect, curvature, distance to drainage network, distance to faults, distance to roads, land cover, lithology, slope, slope length, and TWI—were used based on the availability of the data, geo-environmental conditions of the study area, and literature survey. As a result of the study, it was concluded that the main factor governing the landslides in the study area in all three models is lithology. The artificial factors that trigger landslides across the province of Artvin, as in Şavşat district, are uncontrolled excavation works (usually road widening), uncontrolled explosive excavations, and uncontrolled agricultural land irrigation. In this respect, providing basic disaster awareness trainings to citizens residing in areas susceptible to landslides in the study area and trainings on the causes, effects, and consequences of landslides will be beneficial in terms of risk reduction. Similarly, taking into account landslide susceptibility maps in selecting dwelling zones in rural areas and in determining the routes through which infrastructure facilities such as drinking water, natural gas, electricity, and sewerage will pass, will be effective in reducing the risks associated with landslides in the study area.

Conflict of interest

The authors declare no conflict of interest.

Author details

Halil Akinci[1*], Mustafa Zeybek[2] and Sedat Dogan[3]

1 Department of Geomatics Engineering, Artvin Çoruh University, Artvin, Turkey

2 Güneysınır Vocational School, Selcuk University, Konya, Turkey

3 Department of Geomatics Engineering, Ondokuz Mayis University, Samsun, Turkey

*Address all correspondence to: halil.akinci@artvin.edu.tr

IntechOpen

References

[1] CRED. Natural Disasters 2019. Centre for Research on the Epidemiology of Disasters (CRED). Brussels: CRED; 2020. Available from: https://emdat.be/sites/default/files/adsr_2019.pdf (Accessed: February 2, 2021)

[2] AON. Weather, Climate & Catastrophe Insight: 2020 Annual Report. AON; Chicago, Illinois. Available from: https://www.aon.com/global-weather-catastrophe-natural-disasters-costs-climate-change-2020-annual-report/index.html (Accessed: February 18, 2021)

[3] AON. Weather, Climate & Catastrophe Insight: 2019 Annual Report. AON; Chicago, Illinois. http://thoughtleadership.aon.com/Documents/20200122-if-natcat2020.pdf (Accessed: February 13, 2021)

[4] Varnes DJ. Slope movement types and processes. In: Schuster RL, Krizek RJ, editors. Landslides, Analysis and Control, Special Report 176: Transportation Research Board. Washington, DC: National Academy of Sciences; 1978. p. 11-33.

[5] Rabby YW, Li Y. Landslide susceptibility mapping using integrated methods: A case study in the Chittagong hilly areas, Bangladesh. Geosciences. 2020;10:483. DOI:10.3390/geosciences10120483

[6] Youssef AM, Pourghasemi HR. Landslide susceptibility mapping using machine learning algorithms and comparison of their performance at Abha Basin, Asir region, Saudi Arabia. Geoscience Frontiers. 2021;12:639-655. DOI:10.1016/j.gsf.2020.05.010

[7] Thi Ngo PT, Panahi M, Khosravi K, Ghorbanzadeh O, Kariminejad N, Cerda A, Lee S. Evaluation of deep learning algorithms for national scale landslide susceptibility mapping of Iran. Geoscience Frontiers. 2021;12:505-519. DOI:10.1016/j.gsf.2020.06.013

[8] Yi Y, Zhang Z, Zhang W, Jia H, Zhang J. Landslide susceptibility mapping using multiscale sampling strategy and convolutional neural network: A case study in Jiuzhaigou region. Catena. 2020;195:104851. DOI:10.1016/j.catena.2020.104851

[9] Merghadi A, Yunus AP, Dou J, Whiteley J, Pham BT, Bui DT, Avtar R, Abderrahmane B. Machine learning methods for landslide susceptibility studies: A comparative overview of algorithm performance. Earth-Science Reviews. 2020;207:103225. DOI:10.1016/j.earscirev.2020.103225

[10] Tang Y, Feng F, Guo Z, Feng W, Li Z, Wang J, Sun Q, Ma H, Li Y. Integrating principal component analysis with statistically-based models for analysis of causal factors and landslide susceptibility mapping: A comparative study from the loess plateau area in Shanxi (China). Journal of Cleaner Production. 2020;277:124159. DOI:10.1016/j.jclepro.2020.124159

[11] Trigila A, Iadanza C, Esposito C, Scarascia-Mugnozza G. Comparison of logistic regression and random forests techniques for shallow landslide susceptibility assessment in Giampilieri (NE Sicily, Italy). Geomorphology. 2015;249:119-136. DOI:10.1016/j.geomorph.2015.06.001

[12] Youssef AM, Pourghasemi HR, Pourtaghi ZS, Al-Katheeri MM. Landslide susceptibility mapping using random forest, boosted regression tree, classification and regression tree, and general linear models and comparison of their performance at Wadi Tayyah Basin, Asir region, Saudi Arabia. Landslides. 2016;13(5):839-856. DOI:10.1007/s10346-015-0614-1

[13] Pham BT, Shirzadi A, Tien Bui D, Prakash I, Dholakia MB. A hybrid machine learning ensemble approach based on a radial basis function neural network and rotation forest for landslide susceptibility modeling: A case study in the Himalayan area, India. International Journal of Sediment Research. 2018;33(2):157-170. DOI:10.1016/j.ijsrc.2017.09.008

[14] Wu Y, Ke Y, Chen Z, Liang S, Zhao H, Hong H. Application of alternating decision tree with AdaBoost and bagging ensembles for landslide susceptibility mapping. Catena. 2020;187:104396. DOI:10.1016/j.catena.2019.104396

[15] Luo W, Liu CC. Innovative landslide susceptibility mapping supported by geomorphon and geographical detector methods. Landslides. 2018;15:465-474. DOI:10.1007/s10346-017-0893-9

[16] Chen W, Xie X, Peng J, Wang J, Duan Z, Hong H. GIS-based landslide susceptibility modelling: A comparative assessment of kernel logistic regression, Naïve-Bayes tree, and alternating decision tree models. Geomatics, Natural Hazards and Risk. 2017;8(2):950-973. DOI:10.1080/19475705.2017.1289250

[17] Wang G, Chen X, Chen W. Spatial prediction of landslide susceptibility based on GIS and discriminant functions. ISPRS International Journal of Geo-Information. 2020;9(3):144. DOI:10.3390/ijgi9030144

[18] Kavzoglu T, Sahin EK, Colkesen I. Landslide susceptibility mapping using GIS-based multi-criteria decision analysis, support vector machines, and logistic regression. Landslides. 2014;11(3):425-439. DOI:10.1007/s10346-013-0391-7

[19] Kumar D, Thakur M, Dubey CS, Shukla DP. Landslide susceptibility mapping & prediction using support vector machine for Mandakini River basin, Garhwal Himalaya, India. Geomorphology. 2017;295:115-125. DOI:10.1016/j.geomorph.2017.06.013

[20] Bui DT, Pradhan B, Lofman O, Revhaug I. Landslide susceptibility assessment in Vietnam using support vector machines, decision tree, and Naïve Bayes models. Mathematical Problems in Engineering. 2012;2012:974638. DOI:10.1155/2012/974638

[21] Pradhan B. A comparative study on the predictive ability of the decision tree, support vector machine and neuro-fuzzy models in landslide susceptibility mapping using GIS. Computers and Geosciences. 2013;51:350-365. DOI:10.1016/j.cageo.2012.08.023

[22] Goetz JN, Brenning A, Petschko H, Leopold P. Evaluating machine learning and statistical prediction techniques for landslide susceptibility modeling. Computers and Geosciences. 2015;81:1-11. DOI:10.1016/j.cageo.2015.04.007

[23] Pourghasemi HR, Rahmati O. Prediction of the landslide susceptibility: Which algorithm, which precision? Catena. 2018;162:177-192. DOI:10.1016/j.catena.2017.11.022

[24] Aditian A, Kubota T, Shinohara Y. Comparison of GIS-based landslide susceptibility models using frequency ratio, logistic regression, and artificial neural network in a tertiary region of Ambon, Indonesia. Geomorphology. 2018;318:101-111. DOI:10.1016/j.geomorph.2018.06.006

[25] Sevgen E, Kocaman S, Nefeslioglu HA, Gokceoglu C. A novel performance assessment approach using photogrammetric techniques for landslide susceptibility mapping with logistic regression, ANN and random forest. Sensors. 2019;19(18):3940. DOI:10.3390/s19183940.

[26] Pourghasemi HR, Gayen A, Park S, Lee C-W, Lee S. Assessment of landslide-prone areas and their zonation using logistic regression, LogitBoost, and NaïveBayes machine-learning algorithms. Sustainability. 2018;10:3697. DOI:10.3390/su10103697

[27] He Q, Shahabi H, Shirzadi A, Li S, Chen W, Wang N, Chai H, Bian H, Ma J, Chen Y, Wang X, Chapi K, Bin Ahmad B. Landslide spatial modelling using novel bivariate statistical based Naïve Bayes, RBF classifier, and RBF network machine learning algorithms. Science of the Total Environment. 2019;663:1-15. DOI:10.1016/j.scitotenv.2019.01.329

[28] Hu Q, Zhou Y, Wang S, Wang F. Machine learning and fractal theory models for landslide susceptibility mapping: Case study from the Jinsha River basin. Geomorphology. 2020;351:106975. DOI:10.1016/j.geomorph.2019.106975

[29] Sahin EK. Assessing the predictive capability of ensemble tree methods for landslide susceptibility mapping using XGBoost, gradient boosting machine, and random forest. SN Applied Sciences. 2020;2:1308. DOI:10.1007/s42452-020-3060-1

[30] Sahin EK. Comparative analysis of gradient boosting algorithms for landslide susceptibility mapping. Geocarto International. DOI:10.1080/10106049.2020.1831623.

[31] TURKSTAT. Address based population registration system results. Population of Municipalities, Villages and Quarters. Turkish Statistical Institute. Available from: https://biruni.tuik.gov.tr/medas/?kn=95&locale=tr (Accessed: Mach 2, 2021)

[32] GDM. Meteorological Data Information Sales and Presentation System. General Directorate of Meteorology. Available from: https://mevbis.mgm.gov.tr/mevbis/ui/index.html (Accessed: April 14, 2021)

[33] Keskin I. 1:100,000 Scale Geological Map of Turkey, No:178 Artvin-E48 and F48 Map Sheet. General Directorate of Mineral Research and Exploration, Geological Research Department, Ankara, Turkey, 2013 (in Turkish).

[34] Erener A, Mutlu A, Düzgün HS. A comparative study for landslide susceptibility mapping using GIS-based multi-criteria decision analysis (MCDA), logistic regression (LR) and association rule mining (ARM). Engineering Geology. 2016;203:45-55. DOI:10.1016/j.enggeo.2015.09.007

[35] Parise M. Landslide mapping techniques and their use in the assessment of the landslide hazard. Physics and Chemistry of the Earth, Part C: Solar, Terrestrial and Planetary Science. 2001;26(9):697-703. DOI:10.1016/S1464-1917(01)00069-1

[36] Bera S, Guru B, Ramesh V. Evaluation of landslide susceptibility models: A comparative study on the part of Western Ghat region, India. Remote Sensing Applications: Society and Environment. 2019;13:39-52. DOI:10.1016/j.rsase.2018.10.010

[37] Yanar T, Kocaman S, Gokceoglu C. Use of Mamdani fuzzy algorithm for multi-hazard susceptibility assessment in a developing urban settlement (Mamak, Ankara, Turkey). ISPRS International Journal of Geo-Information. 2020;9(2):114; DOI:10.3390/ijgi9020114

[38] Yan F, Zhang Q, Ye S, Ren B. A novel hybrid approach for landslide susceptibility mapping integrating analytical hierarchy process and normalized frequency ratio methods with the cloud model. Geomorphology. 2019;327:170-187. DOI:10.1016/j.geomorph.2018.10.024

[39] Bahrami S, Rahimzadeh B, Khaleghi S. Analyzing the effects of tectonic and lithology on the occurrence of landslide along Zagros ophiolitic suture: A case study of Sarv-Abad, Kurdistan, Iran. Bulletin of Engineering Geology and the Environment. 2020; 79:1619-1637. DOI:10.1007/s10064-019-01639-3

[40] Akinci H, Kilicoglu C, Dogan S. Random forest-based landslide susceptibility mapping in coastal regions of Artvin, Turkey. ISPRS International Journal of Geo-Information. 2020;9(9):553. DOI:10.3390/ijgi9090553

[41] Soma AS, Kubota T, Mizuno H. Optimization of causative factors using logistic regression and artificial neural network models for landslide susceptibility assessment in Ujung Loe watershed, South Sulawesi Indonesia. Journal of Mountain Science. 2019; 16(2):383-401. DOI:10.1007/s11629-018-4884-7

[42] Pourghasemi HR, Mohammady M, Pradhan B. Landslide susceptibility mapping using index of entropy and conditional probability models in GIS: Safarood Basin, Iran. Catena. 2012;97:71-84. DOI:10.1016/j.catena.2012.05.005

[43] Ba Q, Chen Y, Deng S, Wu Q, Yang J, Zhang J. An improved information value model based on gray clustering for landslide susceptibility mapping. ISPRS International Journal of Geo-Information. 2017;6(1):18. DOI:10.3390/ijgi6010018

[44] Arabameri A, Saha S, Roy J, Chen W, Blaschke T, Bui DT. Landslide susceptibility evaluation and management using different machine learning methods in the Gallicash River watershed, Iran. Remote Sensing. 2020;12(3):475. DOI:10.3390/rs12030475

[45] Youssef AM, Al-Kathery M, Pradhan B. Landslide susceptibility mapping at Al-hasher area, Jizan (Saudi Arabia) using GIS-based frequency ratio and index of entropy models. Geosciences Journal. 2015;19(1):113-134. DOI:10.1007/s12303-014-0032-8

[46] Ding Q, Chen W, Hong H. Application of frequency ratio, weights of evidence and evidential belief function models in landslide susceptibility mapping. Geocarto International. 2017;32(6):619-639. DOI:10.1080/10106049.2016.1165294

[47] Demir G. GIS-based landslide susceptibility mapping for a part of the north Anatolian fault zone between Reşadiye and Koyulhisar (Turkey). Catena. 2019;183:104211. DOI:10.1016/j.catena.2019.104211

[48] Lee S, Choi J. Landslide susceptibility mapping using GIS and the weight-of-evidence model. International Journal of Geographical Information Science. 2004;18(8):789-814. DOI:10.1080/13658811041 0001702003

[49] Akgun A. A comparison of landslide susceptibility maps produced by logistic regression, multi-criteria decision, and likelihood ratio methods: A case study at İzmir, Turkey. Landslides. 2012;9(1):93-106. DOI:10.1007/s10346-011-0283-7

[50] Colkesen I, Sahin EK, Kavzoglu T. Susceptibility mapping of shallow landslides using kernel-based Gaussian process, support vector machines and logistic regression. Journal of African Earth Sciences. 2016;118:53-64. DOI:10.1016/j.jafrearsci.2016.02.019

[51] Chen W, Zhang S, Li R, Shahabi H. Performance evaluation of the GIS-based data mining techniques of best-first decision tree, random forest, and naïve Bayes tree for landslide susceptibility modeling. Science of the Total Environment.

2018;644:1006-1018. DOI:10.1016/j.
scitotenv.2018.06.389.

[52] Gómez H, Kavzoglu T. Assessment
of shallow landslide susceptibility using
artificial neural networks in Jabonosa
River basin, Venezuela. Engineering
Geology. 2005; 78(1-2):11-27.
DOI:10.1016/j.enggeo.2004.10.004

[53] Catani F, Lagomarsino D, Segoni S,
Tofani V. Landslide susceptibility
estimation by random forests technique:
Sensitivity and scaling issues. Natural
Hazards and Earth System Sciences.
2013;13(11):2815-2831. DOI:10.5194/
nhess-13-2815-2013

[54] Akgun A, Erkan O. Landslide
susceptibility mapping by geographical
information system-based multivariate
statistical and deterministic models: in
an artificial reservoir area at northern
Turkey. Arabian Journal of Geoscience.
2016;9:165. DOI:10.1007/
s12517-015-2142-7

[55] Huang F, Cao Z, Guo J, Jiang SH,
Li S, Guo Z. Comparisons of heuristic,
general statistical and machine learning
models for landslide susceptibility
prediction and mapping. Catena.
2020;191:104580. DOI:10.1016/j.
catena.2020.104580

[56] Breiman L. Random forests.
Machine Learning. 2001;45:5-32.

[57] Sun D, Wen H, Wang D, Xu J. A
random forest model of landslide
susceptibility mapping based on
hyperparameter optimization using
Bayes algorithm. Geomorphology.
2020;362:107201. DOI:10.1016/j.
geomorph.2020.107201

[58] Kuhn M. Building predictive models
in R using the caret package. Journal of
Statistical Software. 2008; 28(5):1-26.

[59] Friedman JH. Greedy function
approximation: A gradient boosting
machine. The Annals of Statistics.
2001;29(5):1189-1232.

[60] Pourghasemi HR, Sadhasivam N,
Amiri M, Eskandari S, Santosh M.
Landslide susceptibility assessment and
mapping using state-of-the art machine
learning techniques. Natural Hazards.
2021; DOI:10.1007/s11069-021-04732-7

[61] Wang Z, Liu Q, Liu Y. Mapping
landslide susceptibility using machine
learning algorithms and GIS: A case
study in Shexian County, Anhui
Province, China. Symmetry.
2020;12:1954. DOI:10.3390/
sym12121954

[62] Chen T, Guestrin C. XGBoost. In:
Proceedings of the 22nd ACM SIGKDD
International Conference on Knowledge
Discovery and Data Mining—KDD'16;
New York: ACM Press; 2016. p. 785-794.

[63] Can R, Kocaman S, Gokceoglu C. A
comprehensive assessment of XGBoost
algorithm for landslide susceptibility
mapping in the upper basin of Ataturk
dam, Turkey. Applied Sciences.
2021;11:4993. DOI:10.3390/app11114993

[64] Üstüner M, Abdikan S, Bilgin G,
Balik Şanli F. Crop classification using
light gradient boosting machines.
Turkish Journal of Remote Sensing and
GIS. 2020;1(2):97-105 (in Turkish).

[65] Achour Y, Pourghasemi HR. How do
machine learning techniques help in
increasing accuracy of landslide
susceptibility maps? Geoscience
Frontiers. 2020;11(3): 871-883.
DOI:10.1016/j.gsf.2019.10.001

[66] Zhang Yx, Lan Hx, Li Lp, Wu Ym,
Chen Jh, Tian Nm. Optimizing the
frequency ratio method for landslide
susceptibility assessment: A case study
of the Caiyuan Basin in the southeast
mountainous area of China. Journal of
Mountain Science. 2020;17(2):340-357.
DOI:10.1007/s11629-019-5702-6

[67] Huang Y, Zhao L. Review on
landslide susceptibility mapping using
support vector machines. Catena.

2018;165:520-529. DOI:10.1016/j.
catena.2018.03.003

[68] Tsangaratos P, Ilia I. Comparison of
a logistic regression and Naïve Bayes
classifier in landslide susceptibility
assessments: The influence of models
complexity and training dataset size.
Catena. 2016;145:164-179. DOI:10.1016/j.
catena.2016.06.004

[69] Akinci H, Zeybek M. Comparing
classical statistic and machine learning
models in landslide susceptibility
mapping in Ardanuc (Artvin), Turkey.
Natural Hazards. 2021; DOI:10.1007/
s11069-021-04743-4

[70] Althuwaynee OF, Pradhan B,
Lee S. Application of an evidential
belief function model in landslide
susceptibility mapping. Computers and
Geosciences. 2012;44:120-135.
DOI:10.1016/j.cageo.2012.03.003

[71] Akgun A, Sezer EA, Nefeslioglu HA,
Gokceoglu C, Pradhan B. An easy-to-use
MATLAB program (MamLand) for the
assessment of landslide susceptibility
using a Mamdani fuzzy algorithm.
Computers and Geosciences. 2012;
38:23-34. DOI:10.1016/j.
cageo.2011.04.012

[72] Chen W, Li W, Hou E, Zhao Z,
Deng N, Bai H, Wang D. Landslide
susceptibility mapping based on GIS
and information value model for the
Chencang District of Baoji, China.
Arabian Journal of Geosciences.
2014;7(11):4499-4511. DOI:10.1007/
s12517-014-1369-z

[73] Zhao Y, Wang R, Jiang Y, Liu H,
Wei Z. GIS-based logistic regression for
rainfall-induced landslide susceptibility
mapping under different grid sizes in
Yueqing, Southeastern China.
Engineering Geology. 2019; 259: 105147.
DOI:10.1016/j.enggeo.2019.105147.

[74] Chen W, Sun Z, Zhao X, Lei X,
Shirzadi A, Shahabi H. Performance

evaluation and comparison of bivariate
statistical-based artificial intelligence
algorithms for spatial prediction of
landslides. ISPRS International Journal
of Geo-Information. 2020; 9:696.
DOI:10.3390/ijgi9120696.

[75] Chung CJF, Fabbri AG. Validation
of spatial prediction models for
landslide hazard mapping. Natural
Hazards. 2003;30(3):451-472.
DOI:10.1023/B:NHAZ.0000007172.
62651.2b

Performance Evaluation of Geometric Modification on the Stability of Road Cut Slope Using FE Based Plaxis Software

Fentahun Ayalneh Mekonnen

Abstract

Slope failures are among the common geo-environmental natural hazards in the hilly and mountainous terrain of the world. Specially it is the major difficulty for the development of construction as it causes considerable damage on the infrastructure, human life and property. Different causes of slope failure and stabilization methods are proposed by different scholars. In this study the performance of geometric modification in slope stability was investigated using numerical method. The study uses slope height, slope angle and slope profile i.e. single slope, multi slope and bench slope as a governing parameter in the performance evaluation of geometric modification on the slope stability. The evaluation was conducted on a newly constructed road cut slope using a finite element based plaxis software. The result from performance evaluation of slope profiles show that geometric modification provides better and economical slope stability. The stability of slope decreases with increase in slope height and slope angle leading to an uneconomical design of high slopes in a single slope profile. However, the use of benching improves the stability of cut slope (i.e. the use of 2 m and 3 m bench improves the factor of safety by 7.5% and 12% from single slope profile). The method is more effective in steep slopes. Similarly, the use of a multi slope profile improves the stability of slope in stratified soil with varied strength. The performance is more significant when it is used in combination with benches. The study also provides comparison of slope profiles based on different criteria's and recommend the selection profile based on site-specific considerations.

Keywords: slope profile, bench slope, multi slope

1. Introduction

A slope is an inclined ground surface formed naturally or by excavation for different human activities. Its stability is the major consideration in civil engineering infrastructural projects such as open-pit mining operations, road cut or embankment slopes as its failure causes considerable damage on the infrastructure, human life and property.

Instability of slope can be occurred due to internal or external factors which causes failure either by reducing the shear strength of slope material or by increasing the shear stress on the slope [1]. Different processes such as increased pore pressure, cracking, swelling, decomposition of clayey rock fills, creep under sustained loads, leaching, weathering, and cyclic loading are responsible to a reduction in shear strengths [2]. In contrast to this, the shear stress in slopes may increase due to additional loads at the top of the slope, increase in water pressure, increase in soil weight due to saturation, excavation at the bottom of the slope, and seismic effects [1]. Also, it must be noted that, slope geometry, state of stress, and erosion contributes to the failure of slope. The mechanism of slope failure varies and takes place as speed or slow rate, depending on the type of material, slope geometry, and types of triggering factors. Slide, fall, earth flow, debris flow, topple, planar and wedge failure are the common methods of slope failure.

As both natural and human activities are responsible for the failure of slopes, it is difficult to avoid the problem entirely. However, the level of damage can be significantly reduced by assessing the stability condition and adopting different preventive measures on it. There are various remedial measures applied during and after construction to reduce the impact of slope failure and these can be grouped into four general classes i.e. geometric modification, drainage control, slope reinforcement, and retaining structure [3].

Now a day slope stability studies have been attracted researcher's attention as their understanding on the impact of slope failure in human life and infrastructural development increases. Numerous slope stability studies were carried around the world so far and better understandings are established about causes of failure, mechanisms of failure, methods of analysis, and possible remedial measures. However, the damages due to slope failure are increasing from year to year and still the major difficulty for the development of infrastructural constructions in Ethiopia. A review of previous slope stability assessments [4–6] indicates that slope failures are the main constraint for road and railway construction in Ethiopia. To overcome this problem and acquire better solutions a continuous effort is needed. Hence, this study was carried to evaluate the performance of geometric modification i.e. slope profiles (single slope, multi slope and multi slope) on slope stability using numerical methods.

2. Description of the study area

The study area is located at Adama city in Ethiopia on a newly constructed ring road project. The area is in the east African refit valley system which is dominated by escarpments of various landscapes and bordering. The slope is formed by excavation of volcanic ridges and its height extends up to 40 m with 3 m bench every 10 m. Reddish to brownish color residual soils formed by a physical and chemical process from parent rock and volcanic rocks of different degree of weathering are the major type of materials found in the cut slope (**Figure 1**).

3. Geometry of slope

Geometry is among the most critical factors controlling the stability of the slope [7, 8]. Generally, slope height, slope angle and slope profile are the major parameters in geometric modification. Cut and embankment slopes can be formed using one of the three profiles i.e. single slope, multi slope and bench slope. But depending on the composition of slope material, height of the slope, and hydrological conditions these profiles have different performance in the stability of slope.

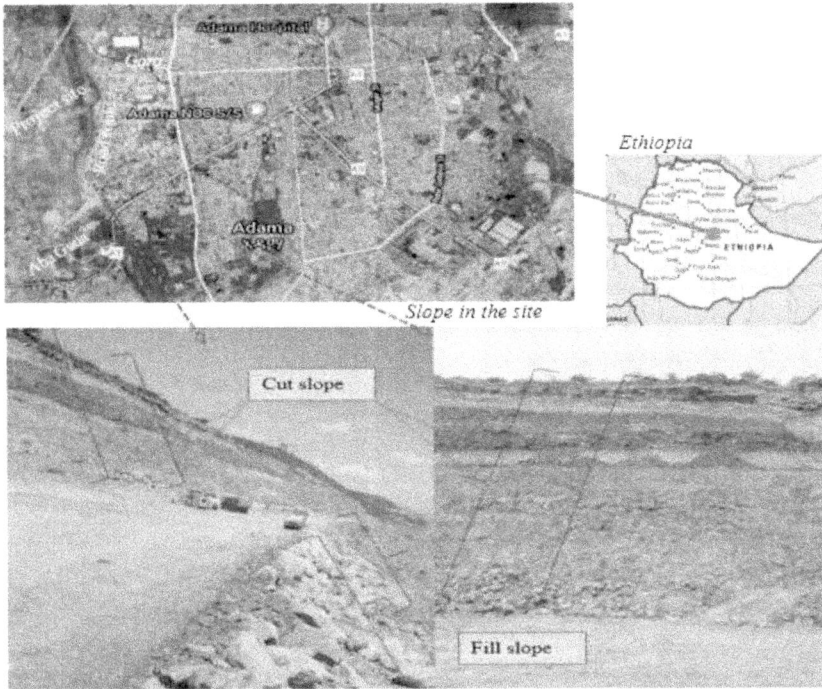

Figure 1.
Location of the study area and slope section of the road.

Single slope profile is used in cut and embankments of dense soils with enough resistance against failure with a limited height [9]. Increasing height (h) and angle of slope (α) will increase the shear stress and decrease normal stress on the potential rupture plane [2]. As a result, h and α are major parameters control the performance of single slope profile.

Multi-sloped profiles are provided in cuts where the stratigraphy of soil consists of two or more layers with different strength characteristics [9]. The method allows the use of both steep and gentle slopes in stiff and weaker layer of the slope section respectively.

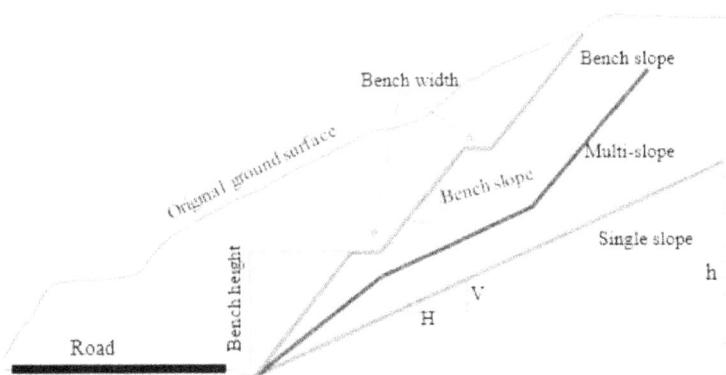

Figure 2.
Geometric profiles of cut slope.

Bench slope profile is a technique in which the overall slope is divided into multiple small slopes. It reduces the driving forces above the failure surface by reducing the weight of slope [8]. Bench slope, bench height and bench width are the major parameters control the performance of bench slopes (**Figure 2**).

4. Methodology

To investigate the performance of slope profiles a newly constructed road cut slope in Adama city was used. Under this investigation the effect of slope profiles and its parameters (slope height, slope angle, bench width, no of bench, and bench angle) on the stability of slope was evaluated interims of FS and deformation. The performance of multi-slope and bench slope profiles was evaluated with respect to single slope profiles and further comparison was made between them interims of construction difficulty, appearance or esthetic value, drainage control, and accessibility for maintenance.

4.1 Numerical modeling

The numerical modeling was carried using finite element method (FEM). The method discretizes a continuum into elements to describe the behavior or actions of individual pieces and reconnecting them to represent the behavior of the continuum [10–12].

5. Material parameters for modeling

For this FE modeling an elastic perfectly plastic Mohr-Coulomb material model was used. The model uses material stiffness (E and v) as elasticity parameter, material strength (φ, and c) as soil plasticity and ψ as angle of dilatancy [10]. The slope section used for this investigation has both soil and rock layers, hence to determine the parameters both field and laboratory tests were carried. The soil shear strength parameters were obtained from direct shear test and its stiffness parameters were correlated with SPT data. Similarly, the strength and deformation parameters of rock layers were determined by correlating field and laboratory tests with rock data software. The geometric and material data used for this investigation were summarized in **Figure 3** and **Table 1**.

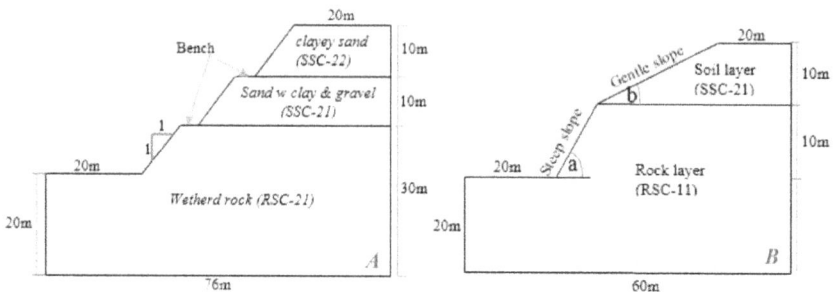

Figure 3.
Cross-section of the newly constructed road cut slope.

Sample code	Description of soil and rocks	γ_{sat} (kN/m³)	γ_{dry} (kN/m³)	C (kN/m²)	$\varphi°$	$\psi°$	E (MPa)	V
SSC-22	Clayey sand	20.1	16.4	14.57	27.89	0	25	0.215
SSC-21	Sand with clay and gravel	20.1	16.6	9.7	35.75	5.75	35	0.261
RSC-11	Slightly to moderately weathered rock	21.3	20.1	130	29.21	0	1962.8	0.35
RSC-12	Highly weathered rock	19.8	18.7	600	20.32	0	686.9	0.38

Table 1.
Material properties used in numerical modeling.

5.1 Stability analysis

To evaluate the stability of slope in different profiles first initial stress and pore water pressure distribution was generated using $k0$ procedure and phreatic water level respectively. Then the deformation and safety analysis were carried with plastic calculation and phi-c reduction method for the same loading conditions. Phi-c reduction is a method where the shear strength parameters (c and tan∅) are successively reduced until the failure occurs [13]. During the process the strength reduction factor ($\sum Msf$) is increased start from 1. The global safety factor is equal to the total multipliers $\sum Msf$ at the point of failure which is expressed as the ratio of initial and reduced strength parameters [10].

$$FS = \sum MSF = \frac{\tan\varnothing_{input}}{\tan\varnothing_{reduced}} = \frac{C_{input}}{C_{reduced}} \tag{1}$$

6. Numerical validation

To validate the numerical model a slope section for this study was evaluated using Fellenius method (analytical solution) and the result was compared with numerical value both in FEM (plaxis) and LEM (slide) software's (**Figure 4**).

Validation also made using slope section first introduced by Zhang [14] and later used by numerous investigators i.e. Ferdlund and Krahn [15], Chen et al. [16], Griffiths and Marquez [17], Zhang et al. [18], and Chaowei et al. [19] to validate their 2D and 3D slope stability evaluations (**Figure 5**). The section was modeled using a slide, Plaxis-2D, and Plaxis-3D software with the same material and boundary condition. The result in **Figure 6** shows a drift of ± 5% from previous investigators. Generally, from both validations the result from numerical modeling shows good agreement with the analytical solution and the previous works.

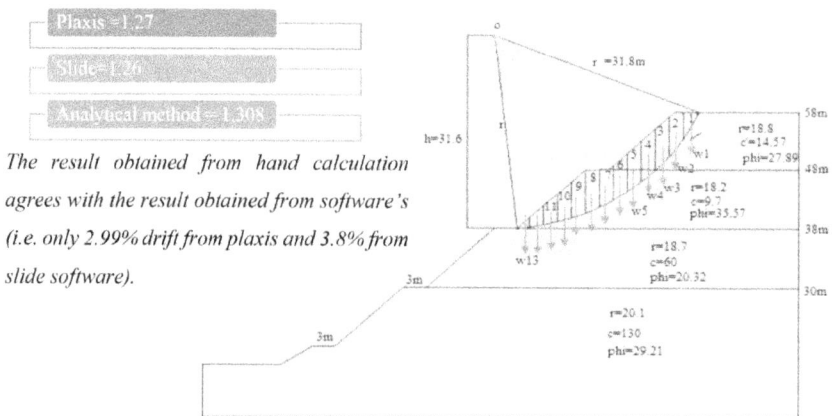

Plaxis =1.27

Slide=1.26

Analytical method = 1.308

The result obtained from hand calculation agrees with the result obtained from software's (i.e. only 2.99% drift from plaxis and 3.8% from slide software).

Figure 4.
FS determination using analytical solutions.

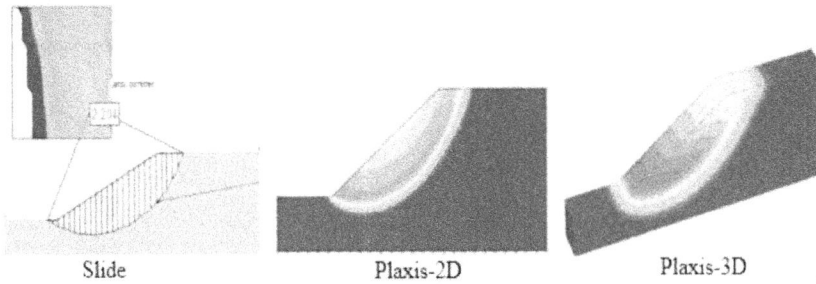

Figure 5.
Modeling of slope section used for validation in slide, Plaxis-2D, and Plaxis-3D.

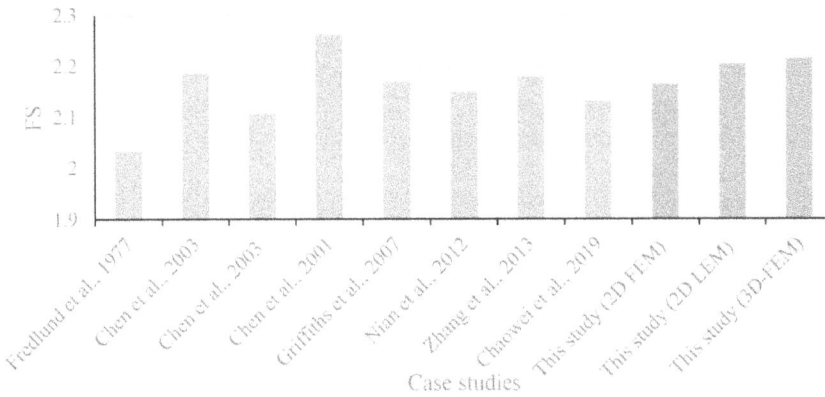

Figure 6.
FS from different researchers and this study on Zhang [14] slope section.

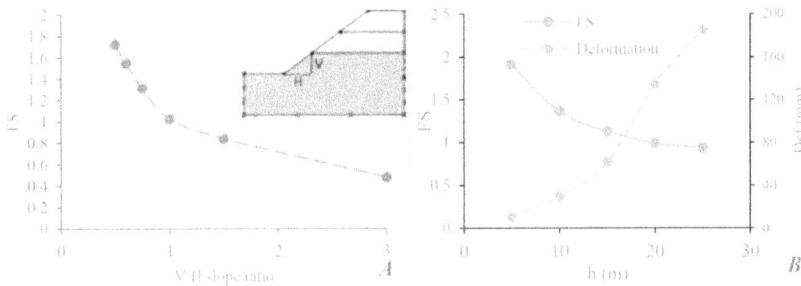

Figure 7.
Effect of slope angle and slope height on the performance of single slope profile.

7. Effect of slope height and slope angle

Increasing angle and height of slope affects the stability by increasing the shear stress and decreasing shear strength on the potential rupture plane. To examine the effect of these parameters numerical models on **Figure 3** was made for different slope height and slope angles.

Figure 7A shows FS of slope for 27°, 34°, 45°, 63°, and 73° (i.e. increasing slope angle reduces the FS). Similarly, **Figure 7B** shows the effect of slope height on FS in an ideally sandy lean clay slope for different slope heights (i.e. increasing slope height increases the deformation and decrease the FS of the slope). Hence it is recognized that both slope height and slope angle reduce the FS of slope in the same principle i.e. increasing self-weight (driving force) above the failure surface and decreasing the normal force on the failure surface.

8. Performance of bench on slope stability

Bench width, cut angle and no bench are the major parameters which control the performance of bench slope profile. To examine the performance of benching and its parameters on the stability of slopes a typical slope section given in **Figure 3** was used. **Figure 8A** shows the FS of slope for 1, 2, 3, 4, 5 and 6 m bench width with a constant 10 m bench height. As the figure indicates stability increases with increasing bench width (i.e. FS increases in 3.5%, 7.7%, and 12% from single slope profile in 1:1 slope ratio for 1 m, 2 m and 3 m bench widths respectively). The percentage change of FS from equivalent single slope profile is 1.7%, 3.6%, 7.7% and 17.4% for 27°, 34°, 45° and 63° respectively in a constant 2 m bench width as shown in **Figure 8B**. Hence the use of benching is more effective in steep slopes.

Similarly, the effect of bench height was evaluated using uniform clayey sand soil in two cases (i.e. case 1, when the overall cut varies with constant slope within bench. Case 2, when overall cut is constant with varied slope within bench as shown in **Figure 8C**). Accordingly, decreasing bench height (increasing no of bench) increases the FS of slope in case 1. However, it has no significant effect in case 2. Generally, benches improve stability of slope in the opposite principle of slope

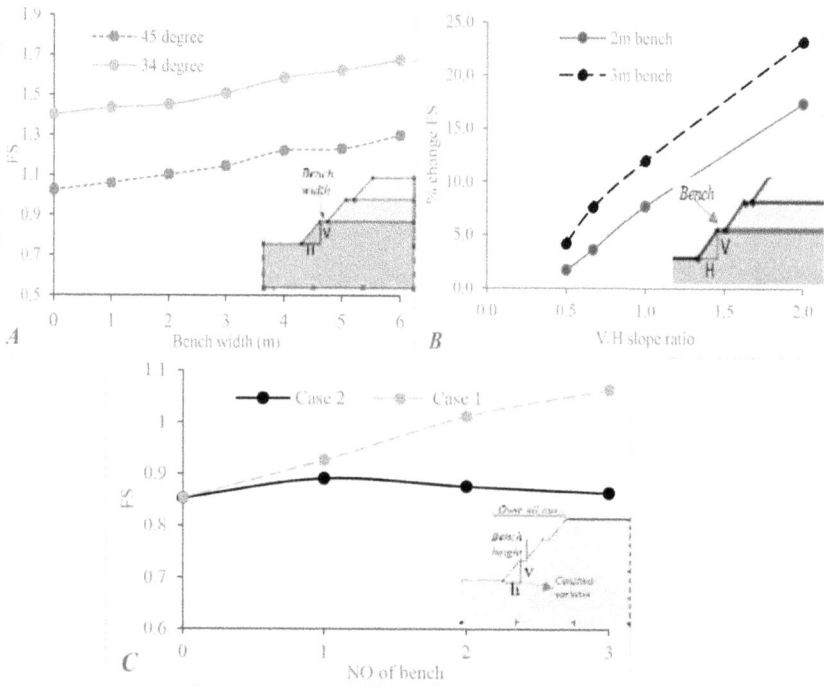

Figure 8.
Effect of bench width, bench slope and bench height on the performance of bench slope profiles.

Slope	a (stronger layer)	45	48	51.3	55	59	63.4	68	78.67	90
	b (weaker layer)	45	42.3	39.8	37.5	35.5	33.69	32	29	26.5
FS		1.202	1.251	1.325	1.414	1.472	1.513	1.55	1.5	1.45

Table 2.
FS for different combination multi-slope profiles (a= stronger slope & b= weaker slope).

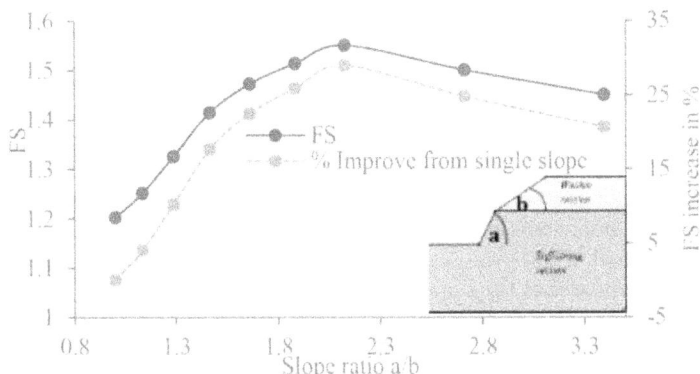

Figure 9.
FS and its change from single slope in % for d/t combination of multi-slope profiles.

height and slope angle by decreasing the driving force of the slope above the failure surface. The method is effective to avoid the use of gentle and high slopes in the design of cut slope.

9. Performance of multi slope profile

To assess the performance of multi slope profile in stratified soils a layered slope section shown in **Figure 3B** is evaluated for different combinations of slope angle (i.e. in the weaker and stronger section). **Table 2** and **Figure 9** shows FS for different combination of weaker and stiffer slope section. Accordingly, FS is improved up to 30% from a single slope profile by adjustment of cut angle (i.e. decreasing weaker section and increasing stiffer section). But it should be reminded that the amount of change in FS depends on the strength characteristics of the slope section. Generally, multi slope profile allows the use of steep and gentle slope in stiff and loses materials respectively in stratified soil.

10. Comparison of slope profiles

Further comparison was made on the performance of profiles by evaluating the slope section in **Figure 10A** for single, multi, bench slope and combination both. **Figure 10B** shows the result of the comparison i.e. FS changes in 13%, 22.7%, and 37.5% from single slope profile in bench, multi-slope, and the combinations of both methods respectively. Hence the use of bench slope, multi slope and combination them provide effective stability in high and stratified slope.

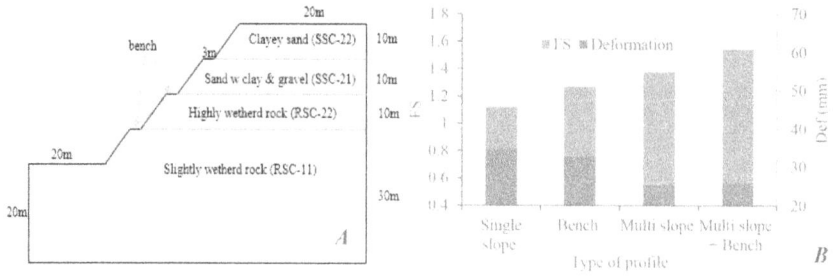

Figure 10.
FS and deformation for different slope profiles.

From the above evaluation, it is recognized that modification of slope geometry is one and the very first economical alternative of slope stability improvement. Although this investigation is made in a specific type of slope material there is no doubt in the role of geometric modification in slope stability. However, the selection of these slope geometry should depend on site-specific parameters i.e. susceptibility of the slope to erosion and infiltration, the variability of slope material, the height of slope, and adjacent area of the slope.

According to the evaluation result, the use of a single slope profile is effective in homogenous stiff slope material when the height of the slope is low. Otherwise, the method may not be safe and economical choice as weak and high slope sections need very gentle slopes. Multi slope profiles are suitable in slopes where there is material strength variability. It provides a very economical slope design without extra excavation by making adjustments only within the slope section. Especially the method is ideal in slopes comprise both rock and soil. The use of bench is an effective geometric measure when the height of the slope is large. It increases FS by reducing the driving force above the failure surface. In addition to this the use of bench slope can provide

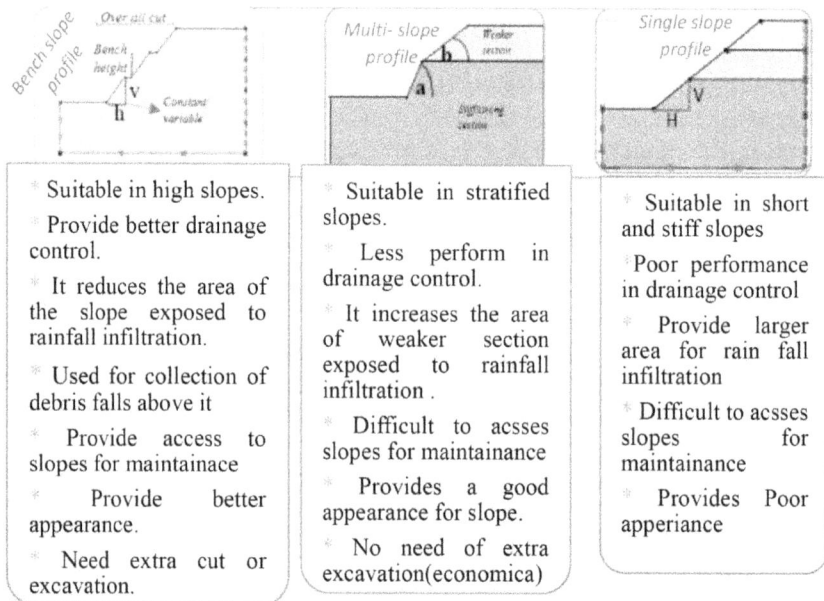

Figure 11.
Advantages and limitations of slope profiles.

the following advantages. (1) It reduces the area of the slope exposed to rainfall infiltration as it allows the use of steep slope between every benching. (2) It provides effective drainage control by collecting the rainwater from each slope profile and draining it laterally to ditches. (3) It provides access to side slopes for maintenance, plantation of vegetation, and decoration. (4) It uses for collection of debris falls above it. (5) It provides esthetic value and better appearance for slopes especially when it located around towns. In general slope profiles have advantages and limitations depending on site specific conditions as shown in **Figure 11**.

11. Conclusion

The performance evaluation of slope profiles in this study was made for the objective of creating awareness on the effect and its suitable condition of different geometric profiles on slope stability. The effect of geometric parameters like slope height, slope angle, bench width, no of bench and bench angle on slope stability were evaluated interims of FS and deformation on selected critical slope sections from the newly constructed road cut slope. From the result, it has been seen that geometric modification will provide better and economical slope stability compared to other structural remedies.

Accordingly, the stability of slope decreases with an increase in slope height and slope angle in single slope profile leading to an uneconomical design of high slopes in a single slope profile. Benching provides an important stability for cut slope especially for slopes having larger height and its performance is more effective in steep slopes. Bench width and bench height also parameters which affect the performance of benching. Multi-slope profile provides an effective slope stability in a stratified soil of varied strength. It allows economical slope design without extra excavation by making adjustments only within the slope section. In addition to its direct effect on the FS, slope profiles have different performance on drainage control, access to maintenance, and its esthetic value. Therefore, the selection of slope profiles during design should be based on site-specific considerations.

Author details

Fentahun Ayalneh Mekonnen
Adama Science and Technology University (ASTU), Adama, Ethiopia

*Address all correspondence to: fantahunayalneh@gmail.com; irccd@astu.edu.et

References

[1] Searom, G. (2017). Landslide Hazard Evaluation and Zonation in and Around Hagereselam Town. A Thesis Submitted to School of Earth Sciences Addis Ababa University, Ethiopia.

[2] Prasad, N. (2017). Landslides – Causes and Mitigation Technical Report Centre for Water Resources Development and Management, India.

[3] Popescu, M. (2001). A Suggested Method for Reporting Landslide Remedial Measures International Union of Geological Sciences Working Group on Landslides, Commission on Landslide Remediation.

[4] Eleyas, A., Li Jian Lin, Costas I. Sachpaz, Deng Hua Feng, Sun Xu Shu, and Anthimos Anastasiadis (2016). Discussion on the analysis, prevention, and mitigation measures of slope instability problems: A case of Ethiopian railways. Electronic Journal of Geotechnical Engineering (21.12), pp. 4101-4119.

[5] Woldearegay, K. (2013). Review of the occurrences and influencing factors of landslides in the highlands of Ethiopia with implications for infrastructural development. Momona Ethiopian Journal of Science 5(1), 3-31.

[6] Samuel, A. (2017). Slope Stability Analysis of Rainfall Induced Landslides. A Case Study on Gohatsion – Dejen Road Abay Gorge. Masters Thesis in Addis Ababa Science and Technology University.

[7] Navya, B., and Hymavathi, J. (2017). Stability Analysis of Slope with Different Soil Types and Its Stabilization Techniques. Department of Civil Engineering, Institute of Aeronautical Engineering, Dundigal.

[8] S. Alfat, L. M. Zulmasri, S. Asfar, and M. S. Rianse. (2019). Slope stability analysis through variational slope geometry using Fellenius method. Journal of Physics: Series 1242.

[9] ERA (2013). Geotechnical Design Manual: Ethiopian Roads Authority Series of Road and Bridge Design Documents. Addis Ababa, Ethiopia.

[10] Burman, A., Acharya, S. P., Sahay, R. R., and Maity, D. (2015). A comparative study of slope stability analysis using traditional limit equilibrium method and finite element method. Asian Journal of Civil Engineering.

[11] Plaxis (2013). Finite Element Program Developed for the Analysis of Deformation, Stability, and Ground Water Flow in Geotechnical Engineering. Ground Water Flow in Geotechnical Engineering Bv, Netherlands.

[12] Raghuvanshi, T. K. (2017). Plane Failure in Rock Slopes – A Review on Stability Analysis Techniques. Journal of King Saud University-Science.

[13] Chen, B. (2018). Finite element strength reduction analysis on slope stability based on ANSYS. Environmental and Earth Sciences Research Journal 4(3), 60-65.

[14] X. Zhang (1988). Three-dimensional stability analysis of concave slopes in plan view. Journal of Geotechnical Engineering 114(6), 658-671.

[15] Fredlund, D.G., and Krahn, J. (1977). Comparison of slope stability methods of analysis. Canadian Geotechnical Journal 14(3):429439. DOI:10.1139/t77-045

[16] Chen, J., Yin, J.-H., and Lee, C.F. (2003). Upper bound limit analysis of slope stability using rigid finite elements

and nonlinear programming. Canadian
Geotechnical Journal 40(4):742752.
DOI:10.1139/t03-032

[17] Griffiths, D., and Marquez, R.
(2007). Three-dimensional slope
stability analysis by elasto-plastic finite
elements. Geotechnique 57(6):537-546.

[18] Zhang, Y., Chen, G., Zheng, L.,
Li, Y., and Zhuang, X. (2013). Effects of
geometries on three dimensional slope
stability. Canadian Geotechnical Journal
50(3):233-249.

[19] Chaowei, S., Junrui, C., Bin, M.,
Tao, L., Ying, G., and Huanfeng, Q.
(2019). Stability Charts for Pseudo-
Static Stability Analysis of 3D
Homogeneous Soil Slopes using
Strength Reduction Finite
Element Method.

Chapter 7

Assessment of Landslide Risk in Ethiopia: Distributions, Causes, and Impacts

Getnet Mewa and Filagot Mengistu

Abstract

The complex geological and geomorphological settings of Ethiopia, consisted of highland plateaus, escarpments, deeply dissected valleys, and flat lowlands, are results of multiple episodes of orogenesis, peneplanation, crustal up-doming, faulting, and emplacement of huge volumes of lava. The broad elevation contrast raging from about −125 m to 4550 m Above Mean Sea Level (AMSL) is an important factor in determining the climate regimes, vegetation types, and even populations' lifestyles. In Ethiopia landslides, mostly manifested as rockfall, earth slide, debris, and mudflow, are among the major geohazard problems that immensely affects life, infrastructures, and the natural environment. They widely occur in the central, S-SW, and N-NW highland regions. This study discusses the distributions, causes, and impacts of landslides and presents a susceptibility zoning map produced applying the weighted overlay analysis method in the ArcGIS environment. For this purpose, key parameters (lithology, elevation, rainfall, slope angel, land use-land cover, and aspect) were selected and assigned weights by considering their contributions to slope failures. Correlations with inventory data have shown very good matching, where more than 90% of the observed data fall in areas categorized either as moderate, high, or very high susceptible zones, where appropriate risk assessments could be mandatory before approval of major projects.

Keywords: orogensis, landslide susceptibility, plateau, rockfall, earth slide

1. Introduction

Landslide is a phenomenon that represents the downward movements of a wide range of slope-forming materials (soils/rocks) due to gravitational and other driving forces [1, 2]. Considering the characteristics of the sliding materials and mechanisms of movements they can be classified as falls, topples, slides, flows, spreads, or any mixture of these and occur either slowly or suddenly. Situated in the horn of Africa between 33 and 48°E longitude and 3.40 and 14.85°N latitude, Ethiopia is the second African nation with a population of about 115 million (www.worldometers.info) and a surface area of 1.122 million km^2. The landscape constitutes highlands plateaus, dissected valleys, escarpments, gentle slopes, and flat plains. These land features are results of geodynamic processes associated with the establishment of the East African Rift System (EARS), which is a narrow

North-west - South-east (NE-SW) elongated rift with thin continental litho-sphere. This rift dissects Ethiopia diagonally into western and eastern plateaus that represent the Nubian and Somalian plates, respectively (**Figure 1**) [3–5]. Active rifting processes combined with local and global drivers (like seismicity, hydrometeorological events, and demographic factors) have created a suitable environment for the widespread effects of landslides. It occurs in the mountain-ous regions of Ethiopia dominantly in the North-Northwest (N-NW), central and South – Southwest (S-SW) highlands, and rift-margins, usually following inten-sive precipitations and brings variable impacts on life, built infrastructures, and natural environment [6–9].

In this work, the distributions, probable causative factors, and impacts of landslides are described with more emphasis on infrastructures using few selected case studies. Applying different secondary sources, a landslide inventory map is compiled and relationships between the natural attributes (lithology, slope height, slope angle, rainfall, and land use-land cover) and spatial distributions of landslides are assessed. Moreover, a susceptibility zoning map is generated involving the mentioned parameters to which weights were assigned considering their signifi-cance to slope failure. Such a map serves as an input to delineate areas according to their importance to various developmental activities and also helps to identify risk

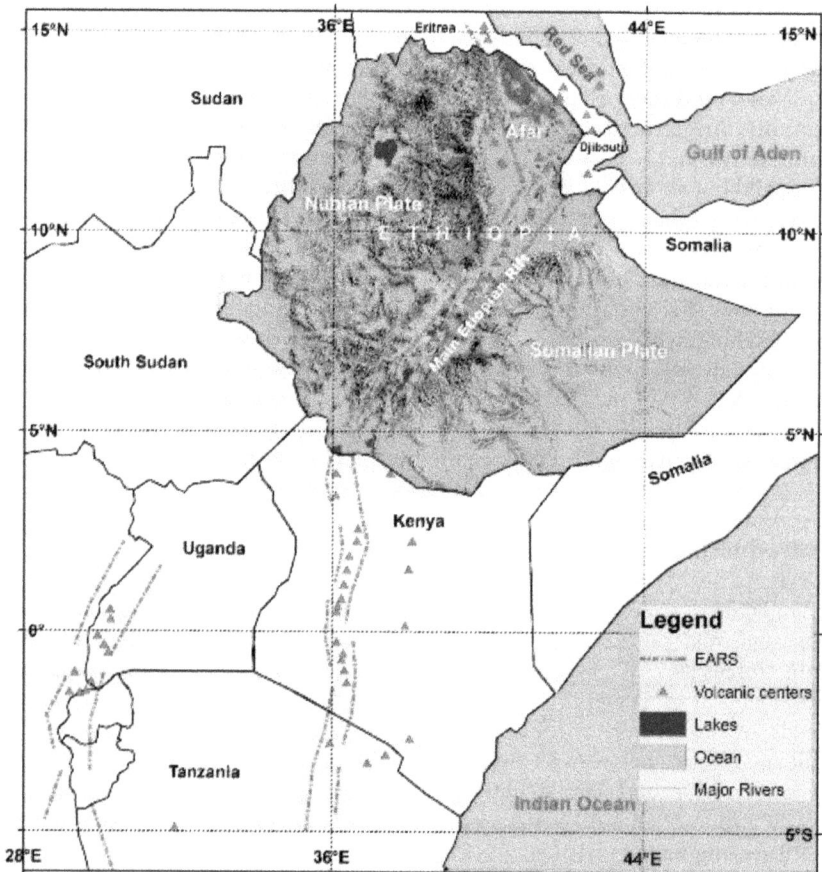

Figure 1.
Generalized map of the East African Rift System (the dotted lines show boundaries of the East African Rift System, while the triangles represent volcanic centers (from Riftvolc consortium, 2013).

potential ones that demand more evaluations and implementation of mitigation measures before major projects are supported.

2. Geomorphology, climate, and general geology

Ethiopia's land surface is characterized by wide elevation contrast that varies from about 125 m below sea level to 4550 m above mean sea level which represents the lowest point in the world, Danakil Depression, and Ras-Dashen mountains (**Figure 2c**). The elevation is the key determinant that defines the climatic conditions of Ethiopia. Accordingly, the country is divided into five climatic zones (**Figure 2a**) that locally known as *Wurch* (very cold), *Dega* (cold), *Weyinadega* (moderate), *Kola* (hot), and *Breha* (very hot temperature zone) [10]. They are distinguished by distinct precipitation and temperature regimes, vegetation and crop types, and even lifestyles of the populations. *Wurch*, *Dega*, and *Weyinadega* climatic zones typically represent the northern, central, and SW highlands as well as rift neighboring plateaus (**Figure 2b** and **c**). They are described by medium-high altitudes (mainly above 1500 m amsl), moderate-high precipitation (above 1000 mm/year), and low-moderate average temperatures (below 25°C). Areas like Tarmaber, Meket, Gashena, Semen Mountains, Arsi, and Bale mountains with elevations above 3200 m amsl that intermittently receive snow and hail (personal communications with local people in 2019; World Institute of Conservation & Environment) constitute this category. Meanwhile, *Kola* and *Bereha* climatic zones, representing the NE (Afar), western (Humera-Metema), S-SW (Gambela, southern Omo), and eastern Ethiopia, show low altitudes, high-very high temperature (above 30–50°C), and very low precipitation (<500, rarely up to 750 mm/year).

The rifting process has defined not only the geomorphology but also the geological settings of Ethiopia, which are discussed in many works [3, 6, 11, 13, 14]. Hence, the formations that underlay the Ethiopian territory differ in composition and age, which ranges from Quaternary to Precambrian (**Figure 2c**). The oldest Precambrian basement rocks are represented by high-grade ortho- and paragneisses and migmatites as well as low-grade volcano-sedimentary—ultramafic assemblages and granitoids [13]. These Precambrian rocks constitute part of the Pan-African Mozambique belt and are distributed in the northern, western, and southern parts of Ethiopia. These formations have undergone prolonged erosion and denudation during Paleozoic that resulted in undulated terrain over which thick Mesozoic sediments (mainly sandstone and limestone) were deposited. The Jurassic sediments cover wide areas of eastern and some places in central and northern Ethiopia. Uplifting of the Afro-Arabian block during Tertiary has resulted in the eruption of a large volume of lava through fractures and covers a substantial part of the country

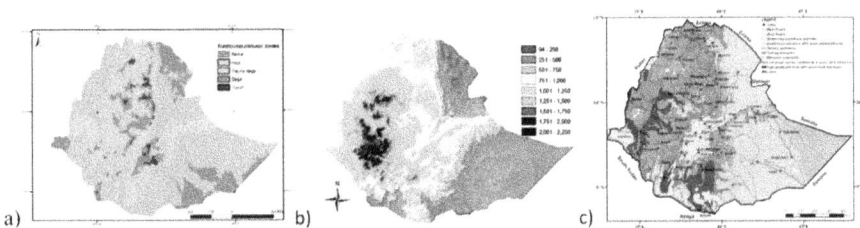

Figure 2.
Climatic zones (a), average annual rainfall distribution (b), and simplified geological (overlain on the topographic) maps of Ethiopia (c). Sources: [10–12].

forming elevated terrains. During this period, sediments deposition took place that cover eastern Ethiopia. Meanwhile, the quaternary period is known for the placement of volcanic lava in areas from Afar depression up to the Lakes Region in the central main Ethiopia rift. Thick Quaternary sediments are distributed in Gambela, Borena, Metema, and few other flat lowland areas (**Figure 2c**).

From the demographic perspective, areas categorized as *Wurch*, *Dega*, and especially *Weynadega* zones, are the most ideal and preferred for settlement due to the availability of sufficient water, fertile lands, and suitable climate for life. But the spread, frequency, and severity of landslides in these areas are more than in *Kola* and *Breha zones*, where the climates are more hostile and flatness of terrain and scarcity of waster do not favor mass movements.

3. Objectives

The basic objective of this study is to examine the distributions, causative factors, and impacts of landslides and acquire a fundamental understanding enabling to develop effective mitigation measures that help to save life and the economy. Accordingly, its specific objectives are: (a) conduct inventory of landslide occurrences across the nation; (b) map links between the spatial distributions and natural attributes that trigger and/or aggravate landslides; (c) assess impacts of landslides on life and infrastructures; d) produce landslide susceptibility zoning map of Ethiopia.

4. Methods and materials

The methodology used in this study comprises—(a) collection and analyses of geological, engineering geological, and geo-hazard data from published and unpublished reports and research publications [11, 15–23]. All data are compiled in the geographic coordinate system using WGS84 datum; (b) collection of rainfall data—the Chirps gridded data for the year 2015 available online was used after comparing it with the National Meteorological Agency (NMA) data, which was found almost alike; (c) download land use-land cover map from National Aeronautics and Space Administration (NASA) web page; (d) data about past landslides events and their impacts. This includes information about the date and time of occurrences, deaths, injuries, forced resettlements, damages to infrastructure, and possible causes; Government offices, non-governmental organizations (NGOs), private firms, research publications, mass media, and local communities, including elder people with knowledge previous events, have served as sources; (e) 30 m resolution DEM data—important inputs about slope height (elevation), slope gradient, and slope direction (aspect) are extracted. These data are closely linked to rainfall and temperature distributions, soil humidity, soli thickness, vegetation types, and density as well as hydrological features of sloppy areas that determine the scale/rates of mass movements; (f) applying a multi-class scoring system based on assigning of weights to selected parameters contributing to slope failure, produce landslide susceptibility zoning map [24, 25].

5. Inventory, distribution, and impacts of landslide

This landslide inventory has identified more than 600 locations across the nation, where landslides occurrences are clearly observed, very few of them are even known with a history of repeated events. Moreover, it reflects localities, where potential landslide risks are imminent [7–9, 15–23, 26–29]. The distribution of inventory data well correlates with lithology, elevation, structural, rainfall, and seismicity maps. Only considering the patterns, landslides occurrences are tentatively classified into four blocks, Block A–D (**Figure 3**). *Block A* represents the N-NE parts of the country, including the eastern part of the western plateau, western rift escarpments, and some places on the rift floor. It stretches from north of Mekele through Michew, Woldiya, Dese, Kombolcha, Kemisse, Shewarobit and continues to the south of Debrebirhan. Major parts of this block are underlain by Tertiary and Quaternary volcanic, whereas Mesozoic sediments are distributed at the NE part of the block covering limited areas of SE Tigray.

Dese and its surrounding are the most well-known areas, where recurrent land-slides cause impacts on settlements, roads, and other properties (**Figure 4a** and **b**). At many places, emerging springs from near surfaces are observed which indicate shallow groundwater. So, steep terrain, undercutting of stream banks, slope erosion, and shallow groundwater are key factors that trigger/aggravate displacement of slope materials. Meanwhile, huge volcanic blocks that are almost detached from the parent rocks are observed at the southern end of the block, in Mushmado village, Say-Debir district, about 8 km from Lemi town (**Figure 4c**). The probability that these blocks would crumble into the valley side is very high if triggered by extreme hydrometeorological, seismic, or other events and will put life, infrastructures, and farmlands in the valley under very high rockfall risk.

Block B encompasses areas between 8 and 13°N latitude and 36.5 and 39°E longitude. Many zones in East and West Gojam (Gozamin, Gonch-Siso Ense, Hulet-Ej-Ense, Shebel-Berenta, Awabel, Aneded, Machakil, Dejen, Adet, Sekela), East Wollega (Ambo, Gedo, Weliso), and Gonder (Lai-Armachoho, Ebinat, Guangua, Quarit) are found here. Moreover, such rivers like Abay, Tekeze, Beshilo, and their main tributaries that formed deep valleys are also among the risk vulnerable areas. The dominant landslide types are rockfall and rock/soil slides, to some extent mudflows. Their impacts on infrastructures and farmlands are quite significant.

The landslides in the Abay gorge, between Dejen and Gohatsion main road, have long and repeated histories, and this economically vital route passes through the 40 km wide Abay (Nile) valley (**Figure 5**). Subsurface investigations carried out within this valley revealed the depths to the slip planes mainly vary are the range of 14–25 m [22]. Even though deaths are not reported, unofficial sources disclosed that the cost of monitoring and road maintenance exceeds 1.5 million USD/year.

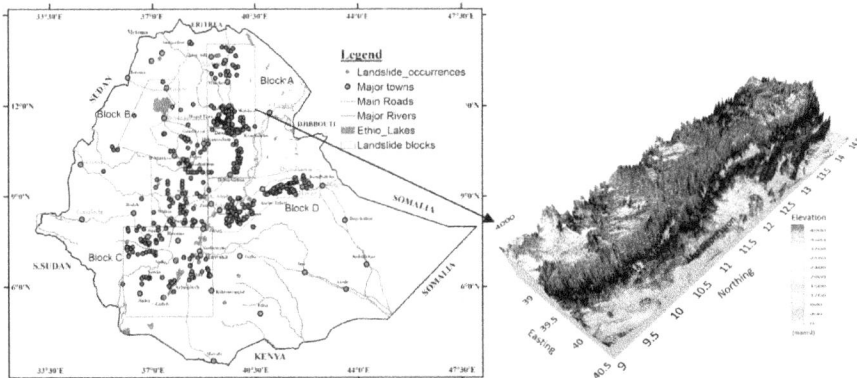

Figure 3.
Landslide inventory map (left) and landscape of NE part of Ethiopia (right).

Figure 4.
Panoramic views of landslides: (a) partial settlement of house foundation, in Dese town; (b) debris slide threatening the Addis Ababa-Dese main road, Kewet district, Debresina town; (c) rockfall risk in Mushmado village, Saya-Debir district, North Shewa zone.

February 2016   June 2010   August 2010   September 2019

Figure 5.
View of landslide occurred in Kurar village, Dejen side (a), the same route, but on the Gohatsion side (b–d): road under maintenance in June 2010 (b), rockfall and debris slide damaged it in August 2010 (c), the site was visited in September 2019 (d).

Block *C* represents south-southwestern Ethiopia and the landslide occurrences are identified within 36–39°E longitude and 5–8°N latitude. It includes Ziway, Shashemene, Hawasa, Hosaina, Adami-Tulu, Jima, Dila, Sodo, Agremariam, Koso Jinka, Sawla, Arbaminch Zuria, Chincha, Gofa, Gidole, Konso, Bako-Gazer, Basketo, and many others places. Along the rift margins, where slope gradients are relatively high, the landslides are manifested by rockfall, debris, and mudflows. A massive landslide that occurred in Gidole, about 55 km SE of Arbaminch town, is a good example that demonstrates how severe is the economic, social, and environmental impacts of landslides in the region (**Figure 6**).

This recent occurrence within the deeply excavated zone (up to 25 m) started in 2009 following intensive rainfalls that saturate the subsurface. The road construction intended to connect Gidole with the Arbaminch-Konso main road has affected the toe parts of the old landslide zone and resulted in the release of shallow groundwater that triggered that landslide. To prevent mass movement slope regarding, about 250 m long retaining walls and drainage ditches were constructed. But due to the large extent of the sliding zone these measures did not change the situation, rather doubled the project cost. So, construction across the failed was abandoned in 2013.

The landslide observed in Alem village, Dodota district, in September 2019 has severely damaged a section on the Dera-Asela main road (**Figure 7a**). The mudflow occurred on May 28, 2018 (**Figure 7a** and **b**) following heavy rainfalls has triggered the sudden movement of a huge volume of earth mass from the head of the landslide and buried houses with 22 people in Western Arsi Zone, Tulu-Gola village, of which 14 were from the same family (May 30, 2018, the Ethiopian reporter).

Block *D* mainly constitutes the eastern part of the Main Ethiopian Rift, such as different districts of East Shewa, Arsi, Harage, Diredawa, and Jigjiga zones. Accordingly, Adama, Chole, Cheleleka, Merti, Fentale, Golelcha, Mechara, Lome, Asebe-Teferi, Bedeno, Kersa, Deder, Chiro, Haromay, Melka-Jilo, Fedis, Gursum, and the areas with landslide records. Rockfall, rock slide, and debris flows are the

Figure 6.
Panoramic view of a landslide body in Welaite village, 2 km NE of Gidole town observed in March 2011 (a), and the same body observed in March 2016 (b). Note that in 2011 its width was about 40 m whereas in 2016 it expanded to about 200 m.

Figure 7.
Road collapse at Alem village, Dodota district, along with the Dera-Assela road (a) and mudslide that killed 22 people and domestic animals in Tulu-Gola village, Western Arsi zone (b and c).

widely observed landslide phenomena. At many places, the landslides are associated with highly weathered and fractured volcanic (ignimbrites and basalts) with steep slope gradients (up to 75°).

In general, this inventory survey has provided tangible information about the spatial distribution, main causative factors, and impacts of landslides. Meanwhile, lack of well-organized records about the types and extents of damages, at this stage it is impossible to give any credible estimations of the economic and environmental losses caused by landslides. Abay A. [30] estimated the losses from 1998 to 2003 to be 135 death, 3500 displaced households, and 1.5 million USD worth of property damages. B. Abebe, et al. [8] stated that landslides that occurred between 1993 and 1998 have claimed hundreds of human lives, damaged over a hundred kilometers of asphalt roads, destroyed many houses, farmlands, and natural vegetations. Similarly, a compilation of data from mass media, newspapers, different reports, and affected communities, (including Fana Broadcasting Corporation; Ethiopian Broadcast Corporation (EBC); Walta Information Center; GSE unpublished technical reports published in 2003–2019) revealed that only between 2016 and 2020 more than 302 people and 1500 domestic animals were killed (**Table 1**).

The landslide in different parts of the country is associated related with three distinct geological setups—(a) landslides developed within the Territory volcanic environment where saturated pyroclastic materials and clay are present as intercalations within the volcanic flows that cover a wide area of the Ethiopian highlands; (b) landslides formed within the sedimentary terrain and the presence of siltstone, shale, and marl as intercalations within the limestone sequence. These are common in the Abay (Nile) valley, in areas south of Mekele (Northern Ethiopia); (c) presence of unstable colluvial materials (silt and clay with gravel and boulder matrix) in areas of relatively gentle terrain covering different formations. Overall, the intercalation within the volcanic and sediments acts as rupture surfaces that aggravate easily displacement of landmasses whenever absorb more fluid in the rainy season.

Region	Landslide affected district (woredas)	Death
Tigray	Hintalo-Wajirat, Hawzen, Atsbi-Wenbera, Degua-Temnbie, Enderta, and Samri-Shart	NR
Amhara	Harbu, Ambassel, Guba-Lafto, Kalu, Dawint, Delanta, Werebabu, Bati, Bugna, Kutaber, Dese-Zuria, Artuma-Farsina, Jille, Efratana-Gidim, Debresina, Kewet, Wagide, Mafud, Mezezo, Chefie-Golana, Dawe-Rahmedo, Gozamin, Gonch-Siso Ense, Hulet-Ej-Ense, Shebel-Berenta, Adet, Sekela, Awabel, Machakil, Dejen, Lai-Armachoho, Ebinat, Guangua, Quarit	11 deaths
Oromiya	Wolmera, Ambo, Guder, Were-Jarso, Kuyu, Jeldu, Tikur, Golelcha, Dodotanasire, Merti, Boset, Aseko, Sude, Dugda-Bora, Wenchi, Welesona Gora, Chela, Chole, Guba-Korcha, Chiro, Dendi, Deder, Kombolcha, Babile, Tullo, Jeju, Daro-Lebbu, Dobba, Seke-Chekorsa, Dedo, Omo-Nada, Goma, Limu-Kosa, Tiro-Afeta, Haromaya, Girawa, Gursum, Chelenko, Bedno, Horo-Guduru	73 deaths and 20 injuries
Southern Nations and Nationalities People (SNNP)	Aleta-Wondo, Kokir Gedebano, Ameya, Gorro, Gumer, Enemorna-ener, Soddo, Meskanena-mareko, Silti, Esara-Tocha, Ela, Marekagena, Decha, Gimbo, Aroresa, Bensa, Dale, Yiga Dera, Shebedino, Yirgachefe, Derashe, Arbaminchzuria, Amaro, Gofazuria, Basketo, Bako-Gazer, Gidole, Konso	102 deaths in one incident
Others	Addis Ababa	116 deaths

Table 1.
Summary of landslide inventory showing affected districts and death and injury reported from 2016 to 2020.

6. Landslide causative factors

The root causes that initiated or accelerated landslide observed at various locations could be associated with the following factors—(a) presence of physically incompetent (soft) earth materials that make up slope surfaces or elevated terrains and also effects of structural discontinuities in areas; (b) intensity and duration of rainfall and effects flooding, erosion a well as groundwater level fluctuations; (c) slope heights and (elevation) and slope angles, which favor mass movements; (d) poor earthwork practices during infrastructure developments (constructions of roads, bridges, dams/reservoirs), and quarrying for mine exploitations. These works involve the removal of earth masses from one place and dumping it into another place which causes either mass deficiency or excess load or both; the effects destabilize slop balances; (e) demographic factor expressed by fast population growth that accompanied by a continuous struggle for resource share. Such struggles put too much pressure on the natural environment and aggravate slope movements; (f) passiveness to enforce code of land-use practices and make accountable those who violate norms; (g) lack of awareness (illiteracy) among rural communities about the influence of landslides in their livelihoods; (h) absence of alternative means of subsistence for rural youth community who have little access to land ownership. So, they rely on over-using of the natural environment that leads to intensive land degradation. Except the natural factors, the human-related ones seem to be fully manageable if better awareness is created, job opportunities are improved and extreme poverty is reduced, land use and land administration codes and practices are enforced, and traditional community practices on land and forest preservations are fully respected. These measures play their role to improve communities' resilience to cope up with the impacts of landslides. The spatial associations between landslide and seismicity are explained in different works [4, 31–33]. In the Ethiopian context, the occurrences of landslides and earthquake epicenters that are practically concentrated within the rift system and surrounding plateaus are found to have very close correlations. But no instrumental records are available that justify the contribution of ground vibrations to triggering landslides.

7. Landslide risk susceptibility zoning

Landslide susceptibility zoning maps are useful tools to differentiate areas that are suitable for agriculture, infrastructure development, national parks, or other purposes as well as delineate risk-prone areas that should be either protected or rehabilitated before approval of any developmental projects [24, 34, 35]. In Ethiopian landslide, mapping and risk zonation were
carried out in specific hazard affected areas, mostly in the highlands and rift regions, using ground survey and remote sensing data [8, 22, 27, 28, 30, 36–40]. However, in this work attempt is made to produce a landslide susceptibility zoning map of the country and correlated with the inventory data acquired through extensive fieldworks mainly by the Geological Survey of Ethiopia, where the lead author has been working for a long time. The field observation data was also used for validation purposes. Thus, the parameters for analyses were selected based on the expert's decision to which weighted values were assigned according to their contributions or influence to slope instabilities [24, 25]. The weights given to involved parameters are as follows: For lithology, elevation, and rainfall—20% each, for slope angle and land use-land cover—15% each, and for aspect—10%. Initially, each of these parameters was sub-divided into five categories, which represent the very low, low, moderate, high, and very high landslide susceptibility zones.

Then using the weighted overlay method in the ArcGIS environment, the map displayed in **Figure 8** is generated. The spatial coverage of each class was calculated by multiplying the corresponding raster counts by the grid pixel sizes and dividing a single class value by the total areal coverage and then multiplying by 100%. Accordingly, about 49.1% of Ethiopia's land surface is susceptible to landslides, of which 39% moderate, 10% high, and 0.1% very high-risk zones. Similarly, 50.9% of the territory is categorized either as very low (5.9%) or low (45%) susceptible zones (**Table 2**).

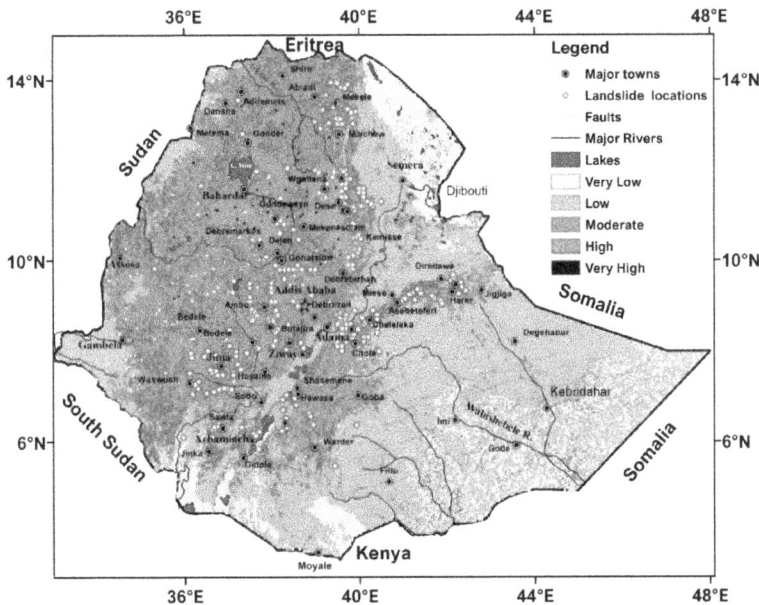

Figure 8.
Landslide susceptibility zoning map of Ethiopia and known landslide occurrences.

No	Susceptible zone	Areal coverage (sq. km)	Country coverage (%)
1	Very low	66,287	5.9
2	Low	504,791	45.0
3	Moderate	437,421	39.0
4	High	112,152	10.0
5	Very high	1448	0.1
	Total coverage	1,122,104	100

Table 2.
Landslide susceptibility zoning.

8. Conclusions

This assessment clearly indicated that landslides are major threats to life, infrastructures, and the natural environment. Natural and human-induced factors (existences of poorly consolidated, easily erodible, saturated and soft earth materials, high slope gradients, intensive or continuous precipitations with subsequent flooding and erosion, scarcity or absence of vegetation cover in sloppy terrains, ground vibrations or seismicity, and continuous growth of population with poor land-use practices) are among the key causes that exposed about 49% of the country to landslide risks. Unfortunately, until the road sector sensed the real challenges posed by a landslide and the ever-increasing rates of fatalities and environmental losses became evident, the issue has never been taken seriously. Hence, it is quite important to proceed with landslide risk assessments to identify and prioritize areas based on their extents, frequency of occurrences, the severity of consequences, as well as nature of different elements exposed to risk. This could be possible through careful considerations of updated landslide inventory data/maps and introducing varieties of risk susceptibility models based on integrated analyses of high-resolution remote sensing and ground observation data, which represent distributions of natural and human-related factors. Ultimately, such comprehensive assessments will play a positive role to ease consequences on life, infrastructures, and the natural environment. It is important to underline that the existing trends of land-use practices are completely inadequate to manage impacts of human-induced landslides that occur very widely. Therefore, implementing zero tolerance for improper land uses through stringent monitoring and enforcement of relevant policies, guidelines, directives, and respecting important social norms must be taken as fundamental tasks of all concerned bodies.

Acknowledgements

We are very grateful to geoscientists of the Geological Survey of Ethiopia (Leta Alemayehu, Habtamu Eshetu, Yewunesh Bekele, Biruk Abel, Abaynesh Mitiku, Tekaligene Tesfaye, Yekoye Bizuye, Debebe Nida, and many others), Addis Ababa University, Ethiopian Roads Authority, National Disaster Risk Management Commission (NDRMC), and other who put tremendous efforts to travel to various parts areas of the country and collect invaluable data used in this assessment. We also extend our sincere appreciation to those who put direct or indirect contributions to this piece of work.

Author details

Getnet Mewa* and Filagot Mengistu
Institute of Geophysics, Space Science and Astronomy, Addis Ababa University, Ethiopia

*Address all correspondence to: getmewa@gmail.com

IntechOpen

References

[1] Varnes DJ. Slope movement: Types and process. In: Schuster RL, Krizek RJ, editors. Landslides: Analysis and Control, Special Report No. 176. Washington D.C.: Transportation Research Board, National Research Council; 1978. pp. 11-33

[2] Hungr O, Leroueil S, Picarelli L. The Varnes classification of landslide types, an update. Landslides. 2014;**11**:167-194

[3] John BS. Synopsis of Geology of Ethiopia. Search and Discovery Article #70215. 2016

[4] Mazzarini F, Keir D, Isola I. Spatial relationship between earthquakes and volcanic vents in the central-northern Main Ethiopian Rift. Journal of Volcanology and Geothermal Research. 2013;**262**:123-133

[5] Chorowicz J. The East African rift system. Journal of African Earth Sciences. 2005;**43**:379-410

[6] Abbate E, Bruni P, Sagri M. Geology of Ethiopia: A review and geomorphological perspectives. In: Landscapes and Landforms of Ethiopia. 2015. pp. 33-64, World Geomorphological Landscapes book Series (WGLC)

[7] Woldearegay K. Review of the occurrences and influencing factors of landslides in the highlands of Ethiopia: With implications for infrastructural development. Momona Ethiopian Journal of Science (MEJS). 2013;**5**(1):3-31

[8] Abebe B et al. Landslides in the Ethiopian highlands and the rift margins. Journal of African Earth Sciences. 2010;**56**(2010):131-138

[9] Ayenew T, Barbieri G. Inventory of landslides and susceptibility mapping in the Dessie area, northern Ethiopia. Engineering Geology. 2004;**77**(1-2): 1-15. DOI: 10.1016/j.enggeo.2004.07.002

[10] Kidanewold BB, Seleshi Y, Melese AM. Surface Water and Groundwater Resources of Ethiopia: Potentials and Challenges of Water Resources Development. 2004

[11] Tefera M, Chernet T, Haro W, Teshome N, Woldie K. Geological Map of Ethiopia. Bulletin/The Federal Democratic Republic of Ethiopia, Ministry of Mines and Energy, Ethiopian Institute of Geological Surveys. No. 3; 1996

[12] Fazzini M, Bisciet C, Billi P. The climate of Ethiopia. In: Landscapes and Landforms of Ethiopia. Edition: World Geomorphological Landscapes. 2010. DOI: 10.1007/978-94-017-8026-1_3

[13] Yibas B et al. The tectonostratigraphy, granitoid geochronology and geological evolution of the Precambrian of southern Ethiopia. Journal of African Earth Sciences. 2002;**34**(2002):57-84

[14] Kazmin V. Geological map of Ethiopia 1:2,000,000, 1st ed. and explanatory notes. Geological Survey of Ethiopia, National Government Publication; 1973

[15] Eshetu H, et al. Engineering Geological & Geohazard Mapping of Dese Map Sheet. GSE Unpublished Technical Report. Addis Ababa: Geological Survey of Ethiopia; 2013

[16] Nida D, Bizuye Y. Geological Hazards and Engineering Geology Map of Hosaina Map Sheet. GSE Unpublished Technical Report. Addis Ababa: Geological Survey of Ethiopia; 2014

[17] Abel B, et al. Akakai Map Sheet Engineering Geological Mapping and Geo-Hazard Assessment. GSE Unpublished Technical Report. Addis Ababa: Geological Survey of Ethiopia; 2012

[18] Eshetu H, et al. Engineering Geological & Geohazard mapping of

Nazareth Map Sheet, GSE Unpublished Technical Report. Addis Ababa: Geological Survey of Ethiopia; 2012

[19] Negash T, Legesse F. Engineering Geological Mapping of Debrebirhan Map Sheet. GSE Unpublished Technical Report. Addis Ababa: Geological Survey of Ethiopia; 2014

[20] Mitiku A. Detail Engineering Geology and Geo-Hazard Investigation of Selected Areas in Jima Map Sheet. GSE Unpublished Technical Report. Addis Ababa: Geological Survey of Ethiopia; 2015

[21] Mitiku A. Engineering Geological and Geo-Hazard Distribution Mapping of Ageremariyam Map Sheet. GSE Unpublished Technical Report. Addis Ababa: Geological Survey of Ethiopia; 2016

[22] JICA-GSE. The Project for Development Countermeasures against Landslides in the Abay Gorge, Ethiopia. Final Report. Addis Ababa: Geological Survey of Ethiopia; 2012

[23] Eshetu H, et al. Geological Hazards and Engineering Geology Maps of Dilla (NB 37-6). GSE Unpublished Technical Report. Addis Ababa: Geological Survey of Ethiopia; 2014

[24] Fell R et al. Guidelines for landslide susceptibility, hazard and risk zoning for land-use planning. Engineering Geology. 2008;**102**(2008):99-111

[25] Soeters R, van Westen CJ. Slope instability recognition analysis and zonation. In: Turner KT, Schuster RL, editors. Landslides: Investigation and Mitigation, Special Report No. 247. Washington DC: Transportation Research Board National Research Council. pp. 129-177

[26] Feseha S, Mewa G. Road failure along the Dedebit-Adiremets road, Northern Ethiopia. Journal of African Earth Sciences. 2016;**118**:65-74

[27] Fubelli G, Guida D, Cestari A, Dramis F. Landslide hazard and risk in the Dessie town area (Ethiopia). In: Presented at the World Landslide Forum, Landslide Science and Practice. Vol. 6. Addis Ababa: Geological Survey of Ethiopia; 2013

[28] Ayalew L. The effect of seasonal rainfall on landslides in the highlands of Ethiopia. Bulletin of Engineering Geology and the Environment. 1999;**58**:9-19

[29] Varilova Z et al. Reactivation of mass movements in Dessie graben, the example of an active landslide area in the Ethiopian Highlands. Landslides. 2015;**12**:985-996. DOI: 10. 1007/s10346-015-0613-2

[30] Abay A, Barbieri G. Landslide susceptibility and causative factors evaluation of the landslide area of Debresina, in the southwestern Afar escarpment, Ethiopia. Journal of Earth Science and Engineering. 2012;**2**(3)

[31] Nowicki Jessee MA, Hamburger MW, Allstadt K, Wald DJ, Robeson SM, Tanyas H, et al. A global empirical model for near-real-time assessment of seismically induced landslides. Journal of Geophysical Research: Earth Surface. 2018;**123**:1835-1859

[32] Tonnellier A, Helmstetter A, Malet J-P, Schmittbuhl J, Corsini A, Joswig M. Seismic monitoring of soft-rock landslides: The Super-Sauze and Valoria case studies. Geophysical Journal International. 2013;**193**:1515-1536

[33] Walter M, Schwaderer U, Joswig M. Seismic monitoring of precursory fracture signals from a destructive rockfall in the Vorarlberg Alps, Austria. Natural Hazards and Earth System Sciences. 2012;**12**:3545-3555

[34] Petschko H, Brenning A, Bell R, Goetz J, Glade T. Assessing the quality of landslide susceptibility maps—case study Lower Austria. Natural Hazards Earth System Science. 2014;**14**:95-118

[35] Regmi NR, Giardino JR, Vitek JD. Modeling susceptibility to landslides using the weight of evidence approach: Western Colorado, USA. Geomorphology. 2010;**115**:172-187. DOI: 10.1016/j.geomorph.2009.10.002

[36] Hamza T, Raghuvanshi TK. GIS based landslide hazard evaluation and zonation: A case from Jeldu district, Central Ethiopia. Journal of King Saud University—Science. 2017;**29**:151-165

[37] Mulatu E, Raghuvanshi TK, Abebe B. Landslide hazard zonation around Gilgel-Gibe-II hydropower project, SW Ethiopia. SINET: Ethiopian Journal of Science. 2009;**32**(1):9-20

[38] Mengistu F, Suryabhagavan KV, Raghuvanshi TK, Lewi E. Landslide hazard zonation and slope instability assessment using optical and InSAR data: A case study from Gidole Town and its surrounding areas Southern Ethiopia. Remote Sensing of Land. 2019;(3):1-14. DOI: 10.21523/gcj1.19030101

[39] Chimidi G, Raghuvanshi TK, Suryabhagavan KV. Landslide Hazard Evaluation and Zonation in and around Gimbi Town, Western Ethiopia—A GIS-Based Statistical Approach. Addis Ababa: Geological Survey of Ethiopia; 2017

[40] Ermias B, Raghuvanshi TK, Abebe B. Landslide hazard zonation (LHZ) around Alemketema Town, North Showa Zone, Central Ethiopia—A GIS based expert evaluation approach. International Journal of Earth Sciences and Engineering. 2017;**10**(01):33-44. DOI: 10.21276/ijee.2017.10.0106

Section 4

Landslide Analysis

Landslide Analysis over Creep Theory - Crack Propagation of Shale Slopes in Şırnak Asphaltite Coal Mine Site 1 and 2

Yildirim İsmail Tosun

Abstract

The soft rock and wet slopes increase landslides over 50 m long creep slide and risk assessment for long steep slide in Şırnak open-pit coal mining should be searched in asphaltite quarries. The Avgamasya quarries No1 and 2 at critical depths and road bench sites in Şırnak, reaching over 120 m height with 60–65° shale slopes, developing major creep factors and other factors for landslide in the deep quarry locations is resulting debris rock falling or free sliding. The pore pressure measurements by measurements of water levels in four wells and water flow counting as the mining safety in recent years. This research provided rock slope stability patterns and crack propagation control of the hazardous location and formation cracks. The stages of creep experimentation explored the geophysical characteristics and thaw and freeze testing of rock samples. For this aim, two different long sliding areas with similar geoseismical conditions, two main analyzing methods, and patterns of researches were developed. Firstly, data on crack propagation *in situ* rock shale faces over certain time periods were determined. Displacement measurements over highly saturated shale—limestone contacts over the base of crack counting in a meter scale such as Rock Quality Designation (RQD) scoring of drilling logs. Secondly, hydrological water level logs were taken into consideration. On the other hand, due to that creep effect over freeze crack propagation unseen cause instability over wet sliding surfaces over 50 m, long sliding surface matter over slopes, poly linear or circle type creep sliding or rock tumbling falling failure types, and GEO5 slope stability, slice analysis will be advantageous instead of Finite Element Method (FEM) method.

Keywords: landslide analysis, Şırnak asphaltite quarry, active potential landslide, creep failure, geotechnical stability, GEO5 slope stability

1. Introduction

The time-dependent failure propagation occurs on the local mountainous natural rockfalls in the hard winter conditions of freezing and thaw cycles observed on road slopes. Hazardous deep quarries in the Şırnak will make a great concern in asphaltite production as significant to the local economy. The hydrology of the area,

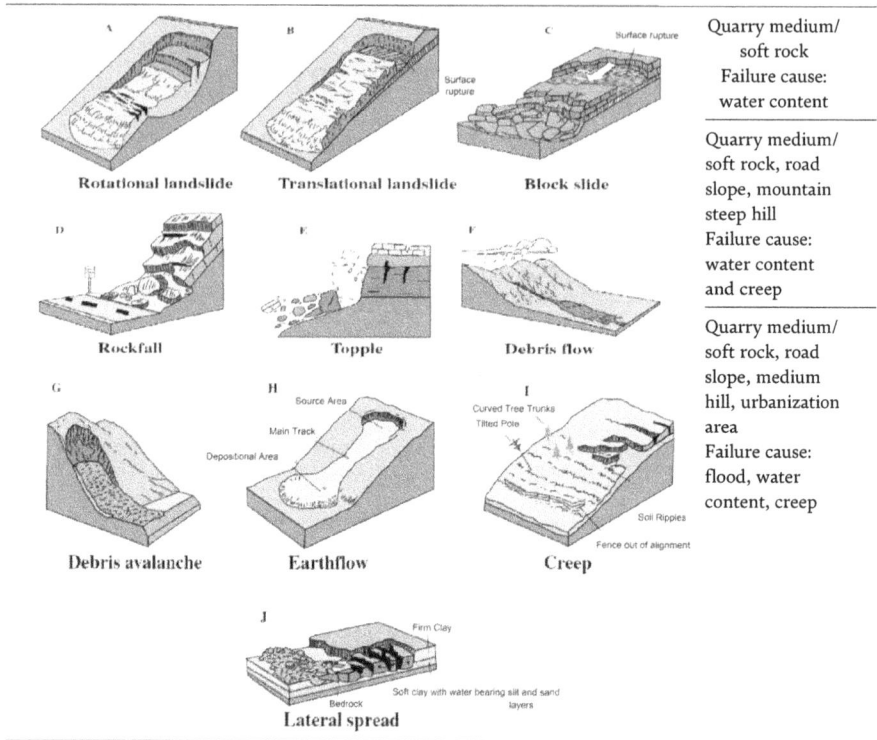

	Quarry medium/ soft rock Failure cause: water content

Rotational landslide   Translational landslide   Block slide

	Quarry medium/ soft rock, road slope, mountain steep hill Failure cause: water content and creep

Rockfall   Topple   Debris flow

	Quarry medium/ soft rock, road slope, medium hill, urbanization area Failure cause: flood, water content, creep

Debris avalanche   Earthflow   Creep

Lateral spread

Table 1.
Creep effect and type of landslides, the sites observed [1].

few months hard fill snow on the quarry avoiding production is important in creep failure or landslides as illustrated in **Table 1** [1] due to loosen rock fallings and free slide of saturated shale slopes over safety limit grade zone [1–3]. Formations such as shale in the regional quarries allow crack propagation by freeze and thaw cycling in the winter climates [4, 5]. The pore structure and low mechanical strength cause a negative effect on creep-dependent breakage quality and stone falling [6–9]. For this reason, the freeze-thaw cycling time and crack texture were critical for creep behavior at the local slope durability [10–12].

Although the $65°$ ($72°$gr) of bench slopes of quarries 1 and 2 as steeper will reduce the excavation costs, it has a negative effect on the creep stability of the quarry at the end of winter, opening the excavation over melted ice period. The highly fractured rock masses have undergone crack propagation, extremely fractured, and showed counting effects on RQD values. The geological strength index GIS, Rock Mass Rating (RMR), and RQD points were determined, by creep texture properties of rock mass in the classification ensured long-term planning stability in the coal quarry excavations [11].

It is quite difficult to creep the block samples depending on the quarry development. Various rock mass classification methods have been proposed. The high groundwater levels and water pressure make ease landslides in the quarry caused major problems in terms of safety. The creep effect in rock mass assessment by freeze and thaw test method is proposed. Q classification system and the Hoek–Brown empirical failure criterion [6–8] were most frequently used by researchers. By the high creep matter, the geomechanical properties critically change the sawing rate resulting in failure by lowering the shear strength and similar methods are used.

The creep failure in rock masses is dependent on discontinuity features that controlled crack face filling and roughness. The slopes failures and discontinuity-controlled failures can be divided into creep discontinuity failures that critically occur in heterogeneous rock conditions as alluvial shale mixed formations. Those creep failures cannot be controlled. The failures are severely fractured and cracked and time depended propagated unseen. It mainly occurs in highly weathered rock masses [6–10].

In the stability analysis, the shear strength of the rock mass at the time of failure was determined. The water pressure parameters for all sliding geometry of the failure surface should be analyzed by the calculated block weight slice method. This method is used in soft rock and heterogeneous rock masses although it describes the failures that occur [1–5], the rock masses are also linear or irregular failure envelopes in different soft rock mass and heterogeneity. However, this cannot fully calculate by the medium the shear strength.

Therefore, The Mohr–Coulomb method is not a preferred measure of instability for rock mass in creep propagation. In the failures that occur in soft and heterogeneous rock masses, Hoek–Brown [6–8] failure criterion is more preferred for the determination of geomechanical strength change.

On quarries no 1 and 2, the south side shale and altered alluvial debris covers and groundwater levels increase in September and reach the highest level in April. December-February period of mining is a closed and active time for creep crack propagation even saturation time [11–14]. The water level increase and freeze and thaw cycle causes of the rock failures occur determined by extensive *in situ* tests. Planning attention to slope geometry in the quarry asphaltite facings contact to water level, drainage, and overburden excavation operations can start in March at the highest water flow of April reaching 50+% filling the bottom of the pit. In order to understand the creep mechanism of slopes S1, S2, S3, and S4 is the main essential issues. While the study is designing the critical hazardous slopes, the geotechnical properties of the rock mass receiving data from the Los Angeles and Blade Sawing tests, freeze-thaw Unaxial Compresssive Strength (UCS) strength is determined [15–20]. Slope angles should be planned considering the quarry safety with a factor count of 1, 35 by GEO5 Slope Stability software. The creep determination process is carried out by freeze-thaw analysis [6].

1.1 Geology in asphaltite quarry in Avgamasya, Şırnak

Study area geology sedimentary alluvial, shale, and calcareous rocks of the Gercus Massif formation Jurassic aged in the Avgamasya, Şırnak province. There are highly disseminated chlorites and calcites are exposed (**Figure 1**). In the southern part, the late Mesozoic aged limestone anticline zone, in the northern part early Eosin age altered porous limestone calcite are located heterogeneous shale contact to Cudi formation and Cizre formation.

In the field studies, the study including the open-pit area has a very heterogeneous layered shale and alluvial contact with vertical asphaltite structure. (**Figure 1**). The hazardous areas of asphaltite quarries are studied as slopes S1, S2, S3, and S4 over the excavation area. The discontinuity intervals were determined. The creep act by freeze and thaw effect is critical for time-dependent rock loose and free landslides developing in mining quarries and urbanization lands in the Southeastern Anatolian regions at height over 1400 m attitudes by high tectonically soft ground conditions [10–14]. The instability of rock loss in the asphaltite coal quarry area creep cracks were developed with advanced mining operations over decades and loosen geotechnical characters of soft heterogeneous formations determined. The detailed investigations in the quarries during mining operations have two fundamental causes of free sliding over freeze and thaw effect on the geotechnical conditions [15–21]. First

Figure 1.
(a and b) View and contour topography of Avgamasya No 1 pit Şırnak asphaltite coal mine site and survey area 1/5000.

of all, the tumbling rock falling landslides occurred at the top of the quarries by groundwater saturation and hard rainwater taking surface conditions as clearly seen. In terms of the past, fatal disasters of instability were observed widely in the different quarries. Secondly, free flow sliding land rocks as debris flows as land flows were similar to the other high deep quarries [22–34]. Therefore, stability conditions and soft rock properties causing past landslides and rock tumbling were so important in order to evaluate and criticize that may develop in the mining excavation areas and even urbanization areas [34–44]. Debris areas or possible free flow loosen landfill areas in mountainous and high-steep rocks were evaluated for free creep flow and tumbling depending on the topology. The unsuitable land use for urbanization over hills increases the creep probability for the development of free land flows [45–53]. In the case of creep landslides, the stability analysis revised by time and related to crack propagation can be achieved and change the safety factor on avoiding the fatal disasters of the quarry or urbanization area concerned [54–60].

The stability analyzes of the top benches in quarries 1 and 2 south side slopes are managed to protect the asphaltite coal excavation equipment and fatal casualties caused by landslides. For this aim, in the quarries 1 and 2 slopes S1, S2, S3, and S4, the free slide top benches three over 35 m long sliding surface excavation area are considered. The fatal experiences of Şırnak Avgamasya and Silopi open-pit mining were carrying high landslide or rock falling risk (**Figure 1**) [11–14]. The creep effect over soft mechanical properties of the soft rock formations of soft limestone, alluvium, and shale layers heterogeneously oriented in the vertical belt form where creep rock falling or free top land flows occurred in the asphaltite quarries. The poly linear surface or circle shape slope stability analyzes for top benches are carried out with GEO5 Slope stability software and GEO5 FEM methods. The slice weight charts of the GEO5 program on the scope of this investigation regarding creep effect, a 1/5000 scale quarry no 1 bench isocontour map covering 3.7 km^2 of the study area are shown in **Figure 1a** and **b**. The high risk of tumbling top rocks and free flow uncohesive sliding over the asphaltite excavation zone is seen as shown in **Figure 2a** and **b**. The blackish zone area is representing a wet asphaltite coal extraction area.

The asphaltite excavation is carried out over 2–4 m thick asphaltite seam placed vertical whirled form in the limestone rock with approximately somehow 1/2 m thick at 62° to SE and approximately 10–25 m for shale 87° to NW and completely changed the orientation to horizontal layer (**Figures 2-4**). The discontinuity surfaces were slightly flat in limestone. It is clear that the crack surfaces are quite slippery in shale rocks. The shale rock mass in the Şırnak quarry pit is extremely fractured (**Figures 3** and **4**). Since it is fractured and heavily weathered over alluvial heterogeneous layers mainly controls free sliding by water saturation and expected collection at the contact surface. In this type of rock formation, landslides and creep failures usually occurred over near-circular failure planes.

2. Method

In the scope of this study, Şırnak asphaltite quarries 1 and 2 in the 940–830 m elevations and 920–810 m elevations. The slope stability analysis for the critical shale slopes were made. The shear stress change corresponding to the creep parameters of rock masses were concluded with tests *in situ* wire extensometers placed. In addition, the RQD and RMR values calculated on the logs as illustrated in **Figure 5** are compared with the values obtained as a result of the freeze-thaw analysis. Later GEO5 stability analysis is carried out to provide operational safety in the quarries in the mine management.

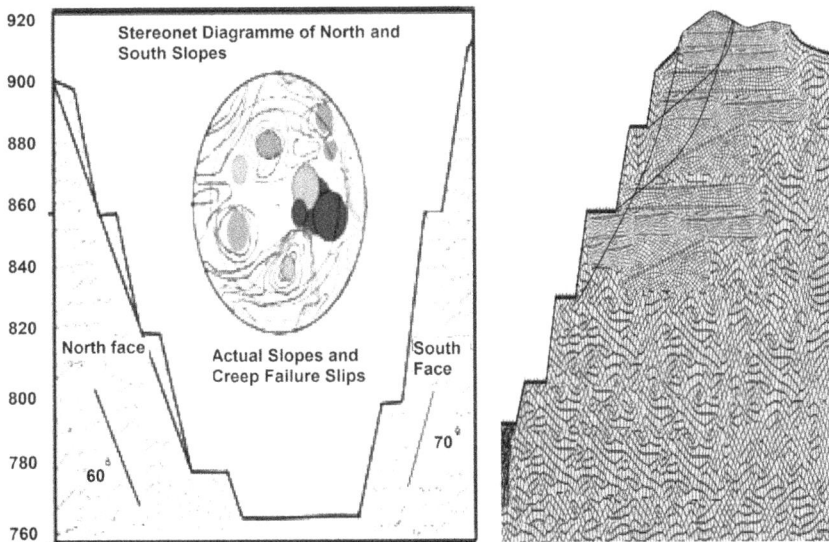

Figure 2.
North and south steep slope face of Avgamasya No 1 pit of Şırnak asphaltite coal mine site and sliding surfaces on a steep slope in the survey area.

Figure 3.
North and south steep slope faces of Avgamasya No 1 and 2 pit survey area.

2.1 Rock mass properties in asphaltite quarry in Şırnak

2.1.1 RMR and RQD

Determination of rock mass properties by RMR method as a result of field studies, RMR and RQD crack counting for shale Jurassic alluvial unit of Pliocene aged are carried out to provide rational stability analysis on creep base regarding two months saturation time cycle. The study area has a lot of facing cracks and cores suitable for determining RQD from the field. RQD value measured as a result of discontinuity in a meter scale line as standard studies is given in **Figure 5**.

RMR score was determined for the determined RQD value and scoring is shown in **Figure 4**. Uniaxial compressive strength UCS and RMR scores of discontinuity

Figure 4.
South steep slope faces of Avgamasya No 1 pit of survey area.

Figure 5.
Shale and alluvium logs of south steep slope faces of Avgamasya No 1 pit survey area.

gap measurements (**Figure 5**) and RMR value and rock classification are presented in **Tables 2** and **3**.

RQD value as scoring for two soft limestones of early Eosins' and Miocene aged in Avgamasya were determined as 45 and 40 scores, respectively. It is concluded that the limestone unit is of medium rock quality and the shale and alluvial unit Pliocene aged is of poor rock quality.

2.2 Pore pressure

The geological rock classification method is useful for slope stability analysis even for complex rock and soil formations. There was a real issue for alluvial pore

Rock formations	Thickness (m)	RQD (%)	c' (kPa)	φ'	P_1 (MPa)	I_1 (MPa) (50 mm)	Shear strength (mm/s)	γsat n (g/cm^3)	γdry (g/cm^3)
S1	25	20.9	700	17	12.0	0.6	34	2.62	2.48
S2	34	22.9	1300	22	15.0	1.1	33	2.65	2.47
S3	35	30.8	1300	23	26.0	1.5	24	2.67	2.52
S4	27	35.9	2700	28	48.0	2.2	14	2.69	2.51

Table 2.
Results from geotechnical tests on samples taken from landslide slopes.

Rock no	S1	S2	S3	S4
γsat max (g/cm^3)	2.92	2.85	2.87	2.67
w$_{opt}$ (%)	15.9	11.9	11.0	12.3
Permeability (k) (cm/s)	5.3×10^{-4}	3.0×10^{-5}	6×10^{-5}	5.3×10^{-4}

Table 3.
Proctor of ground samples and permeability test results.

pressure and rock pore pressure difference and even crack propagation changes the pore pressure in the rock layer put in the calculation. The alluvial soft rock properties are given in **Table 3**.

The pore pressure changing the strength of limestone is illustrated in **Figure 6**.

2.3 UCS compression strength

Samples with volumes 0, 1, 3, and 5% are soaked in water-filled jar. The advantage of this experiment is that it minimizes the errors of the course over 50 mm according to the standard freeze-thaw propagation [15–20]. The UCS change with pore content changing the two limestones, alluvium, and shale in the quarry is illustrated in **Figure 7**.

Considering inferences, extreme deformations can be observed undersaturated with water of pores depending on time. Due to these negative weight effects, various stress changes on complex texture are required for the stable sliding surface in order to reduce cracking and prevent the negative consequences of permeability on the slippery creep. Lower porosities and cracks are seen in two soft limestones. The alluvium and shale reached 22 and 45% cavities by cracking effects of creep.

The pore content of the shale sample containing 30% saturation was determined as 30.5% strength reduction and the maximum dry unit volume weight was 2.85 kN/m^3. Altered limestone reaches a pore saturation of 25% and the maximum dry unit weight of 2.6 kN/m^3 for Şırnak asphaltite quarry (**Figure 8**).

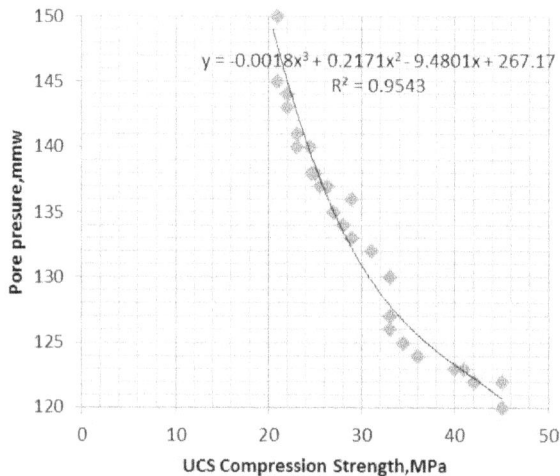

Figure 6.
The UCS compression strength change by pore pressure of soft limestone in Avgamasya asphaltite quarries No 1 and 2.

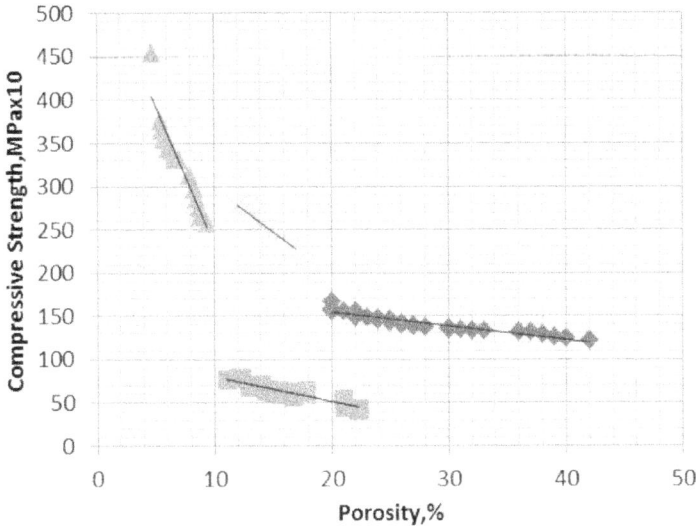

Figure 7.
The UCS compression strength of soft limestones, alluvium, and shale in Avgamasya quarry No 1/2.

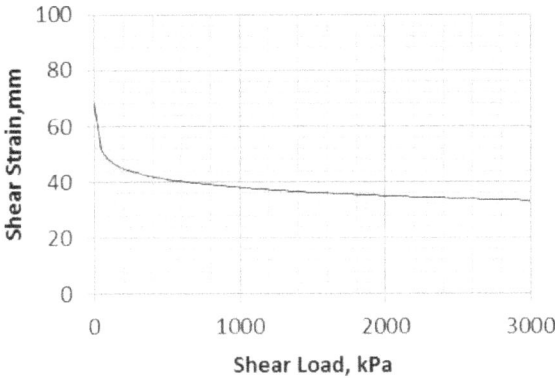

Figure 8.
The shear strain change for Şırnak shale at saturated pore pressure without any creep time for soft limestone.

2.4 Shear strength

The sawing indentation depth for soft and porous rock stones changed by rock microstructure and pore size as given below Eqs. (1) and (2); [16, 17].

Considering filling material by creep moves, extreme deformations can be observed under saturated sliding surfaces depending on time till 35–40 mm scale relative control length at 10 m wire. Due to slippery filling shale mud fines affected uncohesive free slippery surfaces, even internal change on internal friction angle loses caused for the instable sliding surface with increase cracking and prevent the negative consequences of instability on the slippery creep surface.

$$Deformation(cr) = -af \sum_{m=1}^{M} Lcr(1+r)^m \qquad (1)$$

$$E\ Elasticity(cr) = f \sum_{m=1}^{M} Lcr(1+r)^m \tag{2}$$

After this shear process at 70 mm cylindrical disks, the strain amount of the box was determined from the electronic measuring stick on the device. Some samples taken from the submerged part of the box were dried in the oven and the water content corresponding to the pore saturation at the end of the test was found (**Figure 9**).

3. Creep failures of rocks

The stability is provided by water discharge resulted in low deformations of 10–20 mm can be observed under low water pressures of 10 mmw depending on time. The higher weight load of slope slices increases resistive stress change on complex sliding surface texture for the stability with reduction cracking and prevents the negative consequences of permeability on the slippery creep. Eq. (3) shows shear stress with deformation amount θ at time t [30]:

$$u(x; t; \theta) = \sum_{i=0}^{n} u(x, t) + \phi(x; t; \theta).e^{-ti\theta} \tag{3}$$

4. Results and discussions

4.1 GEO5 slope stability analysis on creep theory

GEO5 model weight slice chart construction carried out as given below serial Eqs. (4)–(7) sum [36–41]:

$$F = \sum_{0}^{i} N_i F_i = N_i \frac{C_i}{\gamma H} \cos \beta^i \tag{4}$$

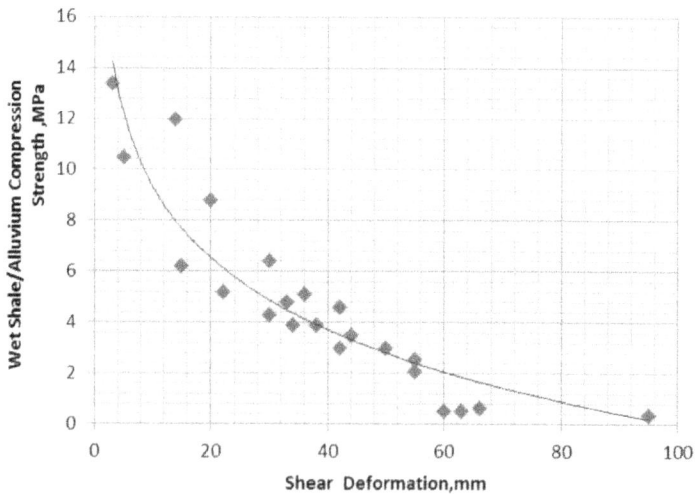

Figure 9.
The shear strength changes by sawing indentation regarding hardness factor of rock depending on creep porosity change by the time.

N_i slice weights the load F_i kN, anisotropic cohesion value of $\frac{C_1}{C_i}$, β free creep slope angle. The free slip surface stability weights show resistance by load chart slice calculations depending on slip surface angle and creep effect. Safety scorings calculated by this resistance to shear should be over 1.35 confirming the stability.

Regarding the crack orientation and intersection with water pore pressures changed by creep (**Figure 6**) in Eq. (5) shear factor R_c varied by slip surface angle exponential rate.

About 2–3 m length slice at i discontinuity at the angle of crack and creep propagated crack density and percentage distribution on slip surface change of $\frac{dy}{dx}$ was calculated by integral as given below Equations 5, 6.

$$R_c = \sum_0^i R_i F_i \tan\theta = \int_a^b e^{-ti\theta} dy^i \tag{5}$$

$$\frac{dy}{dx} = e^{-ti\theta} dy \tag{6}$$

The studied areas shear loads were regressed as exponential functions given below:

The stability mechanism and control by creep crack propagation and creep pore pressure effect for each slice as given in Eq. (7)

$$\frac{dy}{dx} = u = \sum_0^i R_i F_i / \tan a \left(1 - e^{-tRi/\mu}\right)^i \tag{7}$$

u shear deformation by highlighted in the creep theory, the lowered intrinsic friction resistance, F weight slice, a shear fracture inclination angle t time, μ crack free low viscosity at i weight slice.

The safety scoring in toppling and creep flow or landslide is calculated by following the shear force and resisting load over the slope as shear deformations based on the lowered internal friction angle patterns. Rock falling caused by cohesion-free bottom cracking and propagated shear dislocations and pore pressures can be observed in free-fall displacements above 40 mm displacements. The stability analysis carried out by calculations depending on the crack propagation overslip surface for each slice was calculated by the Eqs. (8)–(15) sequentially as below:

$$Ji = \sum_0^i N_i F_i \tan a_i \tag{8}$$

$$Ri = \sum_0^i S_i W_i \cos a_i = \tag{9}$$

$$P_u = \frac{\gamma'}{\gamma} H_i \tag{10}$$

$$Fi_u = \sum_0^i W_i \sin a_i - S_u = \sum_0^i W_i - P_u \tag{11}$$

$$S_{iu} = \sum_0^i c_u' l \ sec \ a_i + Fi_u \frac{\gamma'}{\gamma} H_i \tan^2\phi' \leq 1.25 \tag{12}$$

$$\sigma'_\theta = \sigma - u_a + \chi(u_a - u_{wi}) \tag{13}$$

$$\tau_{\theta i} = c'_i + (\sigma - u_a + \chi(u_a - u_{wi})) \tan\phi' \tag{14}$$

$$S_{iu} = \sum_0^i c_u'l \ sec \ a_i + (\sigma - u_a + \chi(u_a - u_{wi})) \tan^2\phi' \leq 1.25 \tag{15}$$

The safety scoring of S_{iu} water-saturated effective mechanical strength parameters regarding creep failure.

5. Slope analysis of S1, S2 and S3 shale soil/rock face slopes

The top alluvium shale heterogeneous benches of S1, S2, and S3 benches following closed excavation period of winter December and January term started deformation shears at 10 mm sized and the cracks propagated at 11% more and 2 mm widened size gaps caused the little movements that observed and measured in field studies. In quarry No 2, S3 showed free developed slip with top alluvium bench covered with alluvium 10 m sliding depth at steep bench rock stability analyzed by GEO5 programs. The results showed a slip failure problem due to heterogeneous structure and complexity with the wet saturated sliding surface over high 40% saturation on slip surface as given in **Table 4** (**Figures 7** and **10**). In quarry No 1, top shale bench S2 showed similar lowered stability as given in **Table 5** and shown in **Figure 11**. The top bench of soft limestone at slice showed better stability safety factor as given in **Table 6** and illustrated in **Figure 11** as the higher stack and even the maximum height difference between the heel points 30–35 m, the slope of the maximum height of 50 m, the slope of the surface tilt angle is 60°.

The calculation style given in **Tables 4-6** are the results of original wet and creep cohesive resistive parameters obtained from the soft shattered rock formations made in alluvium $c' = 0.9$ kPa, $\varphi' = 20°$, $\gamma_{sat.} = 2.57$ g/cm^3 $c' = 0.4$ kPa, $\varphi' = 21°$, $\gamma_{sat.} =$

Chart	Block height	Block width	Block weight ton	Block weight (kN)	Block shear (MPa)	Resistance to shear (MPa)	Safety	Creep
1	3	4	1.33	13.09	0.77	0.66	1.53	1.32
2	5	6	3.34	32.73	1.62	1.35	1.29	1.08
3	10	8	8.90	87.27	3.97	3.26	1.19	0.98
4	15	9	15.01	147.27	6.57	5.37	1.16	0.95
5	18	7	14.01	137.45	6.14	5.02	1.17	0.95
6	16	5	8.90	87.27	3.97	3.26	1.19	0.98
7	14	4	6.23	61.09	2.84	2.34	1.21	1.00
8	11	4	4.89	48.00	2.28	1.88	1.24	1.03
9	9	3	3.00	29.45	1.47	1.23	1.31	1.09
10	7	3	2.34	22.91	1.19	1.00	1.36	1.14
Total			7.55	666.52	30.82	25.38	1.21	0.99

Table 4.
Weight chart calculations for S1 creep sliding on alluvium.

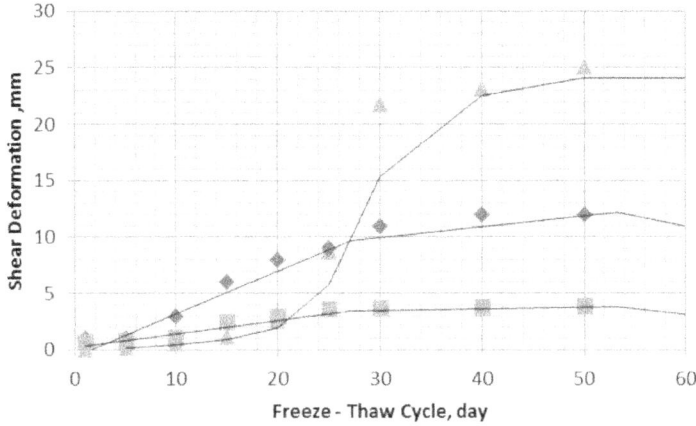

Figure 10.
Creep deformation by time over freeze-thaw time cycle as day periods for soft limestone, alluvium, and shale.

Chart	Block height	Block width	Block weight (ton)	Block weight (kN)	Block chart share (MPa)	Resistance to shear (MPa)	Safety	Creep
1	1.7	2.0	3.34	32.73	1.90	1.62	1.52	1.29
2	2.3	2.3	5.45	53.45	2.97	2.51	1.45	1.23
3	4.0	2.7	10.68	104.72	5.64	4.73	1.41	1.18
4	5.7	3.0	17.01	166.90	8.86	7.42	1.39	1.16
5	6.0	2.3	14.01	137.45	7.33	6.14	1.39	1.17
6	5.7	1.7	9.45	92.72	5.01	4.21	1.41	1.19
7	5.0	1.3	6.67	65.45	3.60	3.03	1.44	1.21
8	4.0	1.3	5.34	52.36	2.92	2.46	1.46	1.23
9	3.0	1.0	3.00	29.45	1.73	1.47	1.53	1.31
10	2.3	1.0	2.34	22.91	1.39	1.19	1.58	1.36
Total			8.59	758.16	41.35	34.79	**1.42**	**1.20**

Table 5.
Weight chart calculations for S2 creep sliding on soft limestone.

2.57 g/cm^3 in shale, and c$'$ = 1.7 kPa, φ' = 25°, $\gamma_{sat.}$ = 2.62 g/cm^3 in soft limestone are used to score safety value. According to calculated safety scores on the potential free creep, surface deformation is lowered to 32° slope as seen in **Figure 11**.

The fracture or discontinuity angle t frequency% in the 20 m sliding on slope direction and the variable position in the design card $\frac{dy}{dx}$ were calculated as give below equations and **Tables 4-6**.

$$R_c = \sum_0^i R_i F_i \tan\theta = \int_a^b e^{-ti\theta} dy^i \tag{16}$$

$$\frac{dy}{dx} = e^{-ti\theta} dy \tag{17}$$

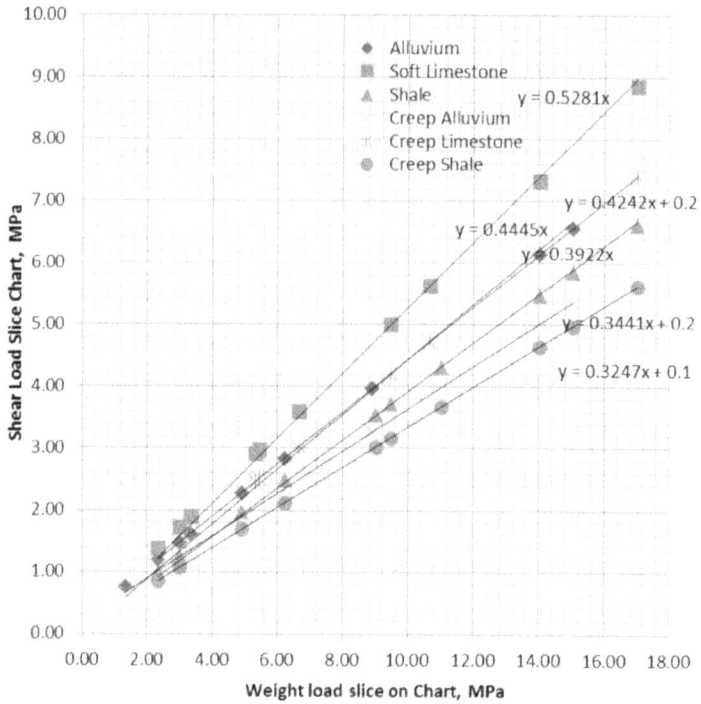

Figure 11.
The shear resistivity on slice weight chart calculations and GEO5 stability analysis for soft limestone, alluvium, and shale with creep.

Chart	Block height	Block width	Block weight (ton)	Block weight (kN)	Block shear (MPa)	Resistance to shear (MPa)	Safety	Creep
1	3.0	3.0	9.01	88.36	3.56	3.03	1.05	0.89
2	3.7	3.0	11.01	108.00	4.32	3.68	1.05	0.89
3	5.0	3.0	15.01	147.27	5.86	4.98	1.04	0.88
4	5.7	3.0	17.01	166.90	6.63	5.63	1.04	0.88
5	6.0	2.3	14.01	137.45	5.48	4.65	1.04	0.88
6	5.7	1.7	9.45	92.72	3.73	3.17	1.05	0.89
7	4.7	1.3	6.23	61.09	2.49	2.12	1.06	0.91
8	3.7	1.3	4.89	48.00	1.98	1.69	1.08	0.92
9	3.0	1.0	3.00	29.45	1.25	1.08	1.11	0.95
10	2.3	1.0	2.34	22.91	1.00	0.86	1.14	0.98
Total			10.22	902.15	36.28	30.86	1.05	0.89

Table 6.
Weight chart calculations for S3 creep sliding on shale.

The safety risk parameter was calculated as 1.42 stable for 40° slopes, but 50° and 60° slopes the safety factors decreased to 1198 and 1060. As given in figure the equation slope 44.2° has given the safety factor for a stable slope as 1120 is shown in **Figure 11**.

$$S = \frac{\sum(c' + \sigma' \cdot \tan \phi') \cdot \ell}{\sum(W \cdot \sin \alpha)} \qquad (18)$$

Data obtained as a result of GEO5 steep discontinuity line studies regarded by shale and alluvial contact layers. The lower internal friction angle and cohesion was giving the safety factor of safety below 1.35 due to a sharp 15–25 m long sliding surface.

In this study, in the estimation of rock mass strength, RQD value for shale and alluvial zone as a result of RQD as an alternative to the method specified forward 41 and 35, respectively (**Table 1**). Use the normal stress(s) value when determining the strengths; m value to be used in GEO5 circle sliding surface analysis on creep base with defeat criterion from 3 for shale, 2 for alluvial clay, and a $35°$ slope are calculated as a function of RMR shale unit its value be 30 RMR crack propagation as determined at the end of creep period for limestone and shale alluvium.

In laboratory experiments for soft limestone rock, uniaxial test strength 42 MPa, dry unit weight 25.8 kN/m³, shale uniaxial compressive strength of the rock 12 MPa, dry unit volume and its weight was determined as 24.45 kN/m³.

2-4 month periods in the winter term results in severe weathering. RQD score is determined as the sum of the cracks and cracks filling points.

6. Conclusions

In this study, two different slope instabilities occurring in the enterprise were investigated. Extremely fractured, fractured, and altered creep saturated units were loaded to software as fill material as sliding saturated creep evaluated by the GEO5 slice analysis method.

The rock mass creep was based on the shear. It was demonstrated using the Hoek–Brown failure criterion. GEO5 FEM analysis was not chosen for the hetero-geneous sliding manner. The shear stress-normal stress graphs gave higher safety values for long sliding surfaces. The slice charts as seen from **Figure 11**, the Avgamasya slopes 1, 2, 3, and 4 in the quarries 1 and 2 for two-dimensional (2D) and three-dimensional (3D) evaluation carried out. The hazard of the sharp slopes in the deep quarry was controlled with the slope stability analyses. The safety coefficients over 1.35 should be considered for steep bench slopes and improved safety working area in bench slopes prepared in Avgamasya asphaltite mining operations. There is also a great creep issue in slope stability is one of the biggest problems. The high pore water saturation comes out of hazardous free sliding. Before instability results arise on slopes various precautions and reinforce-ment methods or appropriate slope design preventing creep failures by applying top alluvial layer geometry, lower slope angles will be necessary for the security of production. For this reason, the displacement on the slopes that are likely to be defeated your angels regularly followed up with monitoring systems.

Finally, the high rainwater conditions or hard and long winter conditions are forced to creep analysis with slope stability charts practiced over the study area.

The 10–40 m long slip surfaces may cause free landslides unexpected in the quarry. The possibility of creep failure land flows may cause fatal accidents for asphaltite coal excavation areas. The stability analysis calculation should be carried out over highly shattered representative specimens at wet saturated effective strengths geotechnical parameters and the results should consider free land flows and tumbling rockslide prevention. The precautious methods appropriate for the asphaltite quarry were water discharge on the site to prevent instability within the scope of the creep in the quarries in the region.

Freeze-thaw cycling in 60 days or two months period decreased the strength values by about 34% over a month period for shale and alluvium. The pore ratio was also similar in the limestone samples. It was increased saturation by 27%. The creep values were also obtained in the shear box strength test.

Creep conditions depending on the pore density in the development of cracks and consequently the formation of wet saturated surfaces and lower shear resistance were observed.

Water discharge over alluvium and shale infiltration of rainwater through the rock slope mass will slow down creep matter and reduce free landslide or flow.

Abbreviations

c'	effective cohesion (kg/cm^2)
c	cohesion (kg/cm^2)
Φ'_0	effective internal friction angle
Φ_0	internal friction angle
τ	shear stress (kg/cm^2)
σ	normal stress (kg/cm^2)
Is	point load index
Bs	bending strength
Ps	compression strength
Wopt	optimum water content
$\gamma_{Natural}$	natural unit volume weight (g/cm^3)
$\gamma_{Saturated}$	saturated unit volume weight (g/cm^3)
γ_{Dry}	dry unit volume weight (g/cm^3)
$\gamma kmax$	maximum dry unit volume weight (g/cm^3)
γs	grain unit volume weight (g/cm^3)
k	permeability coefficient
S1, S2, S3, S4, C1, C2	south and north landslide risk slopes no. 1, 2, 3, 4
S11, C11	sample taken from south and north landslide risk slopes no.

Author details

Yildirim İsmail Tosun
Engineering Faculty, Mining Engineering Department, Şırnak University, Şırnak, Turkey

*Address all correspondence to: yildirimismailtosun@gmail.com

IntechOpen

References

[1] Highland LM. Landslide Types and Processes, U.S. Department of the Interior, U.S. Geological Survey, Fact Sheet 2004-3072. 2004. Available from: https://pubs.usgs.gov/fs/2004/3072/fs-2004-3072.html

[2] Highland LM, Bobrowsky P. The Landslide Handbook—A Guide to Understanding Landslides. Vol. 1325. Reston, Virginia: U.S. Geological Survey Circular; 2008. p. 129. Available from:. https://pubs.usgs.gov/circ/1325/

[3] Hang Lin, Wenwen Zhong, Wei Xiong, Wenyu Tang, 2014. "Slope Stability Analysis Using Limit Equilibrium Method in Nonlinear Criterion", The Scientific World Journal, Hindawi P., vol. 2014, Article ID 206062, 7 pages, https://doi.org/10.1155/2014/206062

[4] Kliche, C. A., 1999. Rock Slope Stability, Society for Mining Metallurgy, SME, Penslyvania, USA, ISBN-10 978 0873351711, 253 p

[5] Kliche, C. A., 2018. Rock Slope Stability, 2nd Edt, Society for Mining, Metallurgy, and Exploration (SME) Penslyvania,USA, ISBN 978-0-87335-369-4

[6] Bieniawski ZT. Engineering Rock Mass Classification. Vol. 1989. New York: Wiley–Interscience; 1989

[7] Bieniawski ZT. Mechanism of brittle failure of rock Part I—Theory of fracture process. International Journal of Rock Mechanics and Mining Sciences. 1967;**4**(4):395-406

[8] Anonymous, 2011. Şırnak Provincial Administration Reports. Şırnak Municipal Bureau, Şırnak, Turkey

[9] Anonymous. Turkey Earthquake Zone Map. Disaster and Emergency Management Bureau, Earthquake Department Ankara; 2011

[10] Anonymous, 2013, GEO5 programs - Engineering Manuals—Part 1–Part 2 http://www.finesoftware.eu/geotechnical-software/, Zaverka Chechk Republik

[11] Anonymous. 2009. GEO5 programs. FEM. Theoretical Guide. Available from: http://www.finesoftware.eu/geotechnical-software/

[12] Tosun Yİ. Asphalt fill strengthening of free slip surfaces of shale slopes in asphaltite open quarry: Stability analysis of free sliding surface for wet shale slopes in Avgamasya Asphaltite Open Quarry No 2. Site, Chapter 5. In: Kanlı AI, editor. Slope Engineering. Rijeka: InTech; 2021. pp. 141-170. DOI:10.5772/intechopen.94893. ISBN: 978-1-83962-924-2, Print ISBN: 978-1-83962-923-5, eBook (PDF) ISBN: 978-1-83962-946-4

[13] Tosun Yİ. Anchorage pile strengthening of shale slopes and cementing falling stone blocks by mixture of melted waste plastics/asphalt and fly ash for slope stability in asphaltite open pit mining site in Avgamasya, Şırnak, Chapter 8. In: Soni A, editor. Mining Techniques—Past, Present and Future. Rejika: InTech; 2021. pp. 141-170. p. 208. DOI: 10.5772/intechopen.69927. ISBN 978-1-83962-369-1

[14] Tosun Yİ. Şırnak ve Cizre Yörel Yerleşim Alanlarındaki Heyelanların Jeoteknik Analizi, Olası Heyelan Tehlike Değerlendirmesi ve Haritalaması, Chapter 3. In: Kiliç GB, Çifçi ÜA, Yilmaz ÜA, editors. Mühendislik Alanında Araştırma ve Değerlendirmeler. Ankara: Gece Kitaplığı, Çankaya; 2019. pp. 35-50. ISBN: 978-605-7631-33-6

[15] ASTM C666/C666M-15. Standard Test Method for Resistance of Concrete

to Rapid Freezing and Thawing. West Conshohocken, PA: ASTM International; 2015. Available from: www.astm.org

[16] ASTM D3080/D3080M-11. Standard Test Method for Direct Shear Test of Soils Under Consolidated Drained Conditions (Withdrawn 2020). West Conshohocken, PA: ASTM International; 2011. Available from: www.astm.org

[17] ASTM D7012-14. Standard Test Methods for Compressive Strength and Elastic Moduli of Intact Rock Core Specimens under Varying States of Stress and Temperatures. West Conshohocken, PA: ASTM International; 2014. Available from: www.astm.org

[18] ASTM D6024/D6024M-16. Standard Test Method for Ball Drop on Controlled Low Strength Material (CLSM) to Determine Suitability for Load Application. West Conshohocken, PA: ASTM International; 2016

[19] ASTM D6067/D6067M-17. Standard Practice for Using the Electronic Piezocone Penetrometer Tests for Environmental Site Characterization and Estimation of Hydraulic Conductivity. West Conshohocken, PA: ASTM International; 2017

[20] ASTM D5878-19. Standard Guides for Using Rock-Mass Classification Systems for Engineering Purposes. West Conshohocken, PA: ASTM International; 2019

[21] Gorzelanczyk T, Schabowicz K. Effect of freeze–thaw cycling on the failure of fibre-cement boards, assessed using acoustic emission method and artificial neural network. Materials (Basel). 2019;**12**(13):2181. DOI: 10.3390/ma12132181

[22] Hoek E, Bray JW. Rock Slope Engineering. Hertford: Stephen Austin and Sons, Ltd; 1977. p. 402

[23] Lamp WT, Whitman RV. Soil Mechanics. New York: John Wiley and Sons; 1969

[24] Bishop AW. The use of the slip circle in the stability analysis of earth slopes. Geotechnique. 1955;**5**:7-17

[25] Hoek E. Estimating the stability of excavated slopes in Opencast mines. Institution of Mining and Metallurgy. 1970;**A105**:A132

[26] Paşamehmetoğlu LV, Özgenoğlu A, Watermelon C. Rock Slope Stability. 2nd ed. T.M.M.O.B Mining Eng: Bureaou Publications, Ankara; 1991

[27] Anbalagan R. Landslide hazard evaluation and zonation mapping in mountainous terrain. Engineering Geology. 1992;**32**:269-277

[28] Görög P, Török A. Slope stability assessment of weathered clay by using field data and computer modeling: a case study from Budapest. Natural Hazards and Earth System Sciences. 2007;**7**:417-422. Available from: www.natu-hazards-earth-syst-sci.net

[29] Görög P. Stability Problems of Abandoned Clay Pits in Budapest, IAEG 2006, Paper Number 295. The Geological Society of London; 2006

[30] Dramis F, Sorriso-Valvo M. Deep-seated gravitational slope deformations, related landslides and tectonics. Engineering Geology. 1994;**38**:231-243

[31] Hoek E. Practical Rock Engineering, notes by Evert Hoek Hoek. 2013. Available from: http://www.rocscience.co

[32] Hutchinson JN. Landslide hazard assessment. Keynote paper. In: Bell DH, editor. Landslides, Proceedings of 6th International Symposium on landslides, Christchurch, New Zealand. Vol. 1. Rotterdam: Balkema; 1995. pp. 1805-1841

[33] Prusa J. In: Vanicek et al., editors. Comparison of geotechnic softwares - Geo FEM, Plaxis, Z-Soil, XIII to ECSMG. Vol. 2. Prague. ISBN 80-86769-01-1: cgts; 2009

[34] Vaneckov V, Laura J, J Prus, 2011. Sheeting Wall Analysis by the Method of Dependent Pressures, Geotec Hanoi Geotec Hanoi 2011 "Geotechnics for sustainable development:"- ISBN 978-604-82-000-8 ID No. / pp. 7 Vietnam

[35] Venkatramaiah, C., 1993. "Stability of Earth Slopes" "Geotechnical Engineering", New Age Int. Pub., Trubati, India

[36] Tosun Yİ, Cevizci H, Ceylan H. Landfill Design for Reclamation of Şırnak Coal Mine Dumps—Shalefill Stability and Risk Assessment, ICMEMT 2014, 11-12 July 2014. Chekoslovakia: Prag; 2014

[37] Tosun Yİ. A case study on use of foam concrete landfill on landslide hazardous area in Şırnak City Province. In: XX Congress of the Carpathian Balkan Geological Association; 24–26 September 2014; Tirana, Albania. 2014

[38] Tosun Yİ. Shale stone and fly ash landfill use in land-slide hazardous area in Sirnak City with Foam Concrete. GM Geomaterials Journal. 2014;4(4): 141-150. DOI: 10.4236/gm.2014.44014

[39] Yıldırım İ. Tosun, 2016, Kalker, Marn ve Şeylin Sünme Karakterizasyonu - Bitümlü Gözenekli Agrega için Don—Mikrodalga Kurutma-Bilya Darbe Dayanım Testi ile Sünme Etüdü, AGGRE 2016, 8th Internatıonal Aggregates Symposıum; October 5–7; Istanbul, Turkey

[40] Tosun, Y.İ., 2016. Use of Modified Freeze-Drop Ball Test for Investigation the Crack Propagation Rate in Coal Mining- Case Study for the Şırnak Asphaltite Shale, Marly Shale and Marl in Şırnak Coal Site, IBSMTS 2016, 1th Internatıonal Black Sea Symposium on Mininig and Tunnelling, November 02–04; Trabzon, Turkey

[41] Török, Á. Bögöly, G., Czinder, B., Görög, P., Kleb, B., Vásárhelyi, B., Lovas, T., Barsi, Á., Molnár, B., Koppányi Z., and Somogyi, J. Á., 2016. Terrestrial laser scanner aided survey and stability analyses of rhyolite tuff cliff faces with potential rock-fall hazards, an example from Hungary, Eurock, Cappadicia, 877–881.

[42] Á Török, Á Barsi, G Bögöly, T Lovas, Á Somogyi, and P Görög, 2017. Slope stability and rock fall hazard assessment of volcanic tuffs using RPAS and TLS with 2D FEM slope modelling, Nat. Hazards Earth Syst. Sci. Discuss., Nat. Hazards Earth Syst. Sci., 18, pp. 583–597, doi:10.5194/nhess-2017-56, 2017

[43] Barton N, Lien R, Lunde J. Engineering classification of rock masses for the design of tunnel support. Rock Mechanics. 1974;6(4):189-236

[44] Bieniawski ZT. Engineering Rock Mass Classifications: A Complete Manual for Engineers and Geologists in Mining, Civil, and Petroleum Engineering. Hoboken, New Jersey: John Wiley & Sons; 1989

[45] Cai M, Kaiser PK, Tasaka Y, Minami M. Determination of residual strength parameters of jointed rock masses using the GSI system. International Journal of Rock Mechanics and Mining Sciences. 2007;44(2): 247-265

[46] Dong-ping D, Liang L, Jian-feng W, Lian-heng Z. Limit equilibrium method for rock slope stability analysis by using the Generalized Hoek–Brown criterion. International Journal of Rock Mechanics and Mining Sciences. 2016;89(June): 176-184

[47] Erguler ZA, Karakuş H, Ediz IG, Şensöğüt C. Assessment of design

parameters and the slope stability analysis of weak clay-bearing rock masses and associated spoil piles at Tunçbilek basin. Arabian Journal of Geosciences. 2020;**13**(1):1-11

[48] Hammah, R., Curran H, J., Yacoub, T., Corkum, B. 2004. Stability analysis of rock slopes using the finite element method. In Proceedings of the ISRM Regional Symposium EUROCK 2004 and the 53rd Geomechanics Colloquy, Salzburg, Austria

[49] Hoek E, Carranza C, Corkum B. Hoek-brown failure criterion. 2002 edition. Narms-Tac. 2002;**1**(1):267-273

[50] Hoek E, Kaiser PK, Bawden WF. Support of underground excavations in hard rock. Journal of Rock Mechanics and Mining Sciences. 2000;**35**(2):219-233

[51] Hoek E. Estimating Mohr-Coulomb friction and cohesion values from the Hoek-Brown failure criterion. International Journal of Rock Mechanics and Mining Sciences. 1990;**27**(3): 227-229

[52] Sheng H, Kaiwen X, Feng D. Establishment of a Dynamic Mohr–Coulomb Failure Criterion for Rocks. International Journal of Nonlinear Sciences and Numerical Simulation. 2012;**13**(2012):55-60

[53] Park J, Hyun C-U, Park H-D. Changes in microstructure and physical properties of rocks caused by artificial freeze–thaw action. Bulletin of Engineering Geology and the Environment. 2015;**74**(2):555-565

[54] Ning L, Jonathan G. Infinite slope stability under steady unsaturated seepage conditions. Water Resources Research. 2008;**44**:11404. DOI: 10.1029/2008WR006976

[55] Duncan JM, Wright SG. Soil Strength and Slope Stability. Hoboken, New Jersey: John Wiley; 2005. p. 297

[56] Hoboken NJ, Fredlund DG, Morgenstern NR, Widger RA. The shear strength of unsaturated soil. Canadian Geotechnical Journal. 1978;**15**:313-321

[57] Rahardjo HT, Ong H, Rezaur B, Leong EC. Factors controlling instability of homogeneous soil slopes under rainfall. Journal of Geotechnical & Geoenvironmental Engineering. 2007; **133**(12):1532-1543

[58] Bazant ZP et al. In: Bazant Z, editor. Chapter 3 Creep Analysis of Structures, Mathematical Modeling of Creep and Shrinkage of Concrete. Hoboken, New Jersey: John Wiley & Sons Ltd; 1988

[59] Hoek, E., Carranza, C., Corkum, B. 2002. Hoek-brown failure criterion – 2002 edition. 2002 Edition. Proceedings of the 5th North American Rock Mechanics Symposium, Toronto, 7-10 July 2002, 267-273.,Narms-Tac, **1**(1), 267–273.

[60] Wei Y, Jiaxin L, Zonghong L, Wei W, Xiaoyun S. A method based on the Generalized Hoek-Brown (GHB) criterion for strength reduction slope failure (Iran). Environmental Earth Sciences. 2020;**60**(1):183-192

Chapter 9

Analysis of Landslide and Land Subsident Using Geophysical Method in the East Java Province, Indonesia

Adi Susilo, Sunaryo Sunaryo, Eko Andi Suryo,
Turniningtyas Rachmawati
and Muwardi Sutasoma

Abstract

East Java Province, which is geologically very complex, often occurs natural disasters, especially landslide and land subsidence. The area of East Java is divided into 3 parts, namely the southern part which is the result of volcanic lahar, and also the uplift from the southern sea. Those two kinds of sediment, geologically is quarter and tertiary volcanic deposits age, and limestone. The Middle part, is a cluster of active volcanoes that are quarter old, which provide quarter-aged sediments and these area is rich in geothermal. The Northern part, which is a sediment from the Java Sea and the Madura Strait, with several limestone mountains, is an area rich in hydrocarbons. The area to be studied is the Southern area, namely the quarter sediment from volcanic lava and the lifting of limestone which has the potential to occur landslides and land subsident. The landslide and land subsident symptoms will be analyzed using the geophysical method, to predict the landslide volume and also the dangerous areas with regard to the land subsident.

Keywords: Sediment, Landslide, land subsident, Geophysical Method, Indonesia

1. Introduction

Indonesia is a country that often experiences hydrometereological disasters, including disasters caused by climate change and weather [1]. The National Disaster Management Agency (BNPB) noted that, in the period of 2020, Indonesia has experienced 2,925 natural disasters, starting from Wednesday (1/1/2020) to Tuesday (28/12/2020). According to data compiled by BNPB, the disasters that occurred throughout 2020 were dominated by hydrometeorological natural disasters such as floods, flash floods, landslides, hurricanes, droughts, forest and land fires. Based on the details of hydrometeorological disaster data, flood events have occurred up to 1,065 events throughout 2020. Disasters caused by hurricanes have occurred as many as 873, landslides as many as 572 events. Furthermore, for forest fires there have been 326 incidents, tidal waves and abrasion have occurred 36 events, and droughts have occurred as many as 29 events. For the types of geological and volcanic disasters,

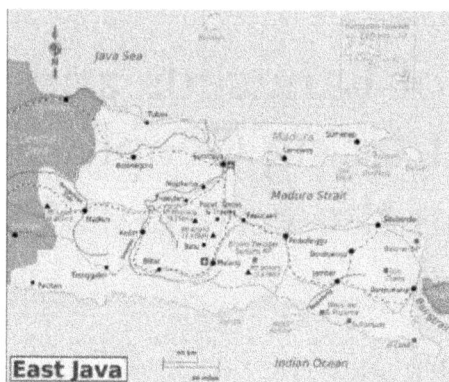

Figure 1.
Map of East Java [2].

there are 16 and 7 events, respectively. There were 370 people who died as a result of the natural disaster, 39 people were missing and 536 people were injured.

East Java, is one of the provinces in Indonesia, which is on the island of Java. This province is located in the west of the province of Bali and in the east of the province of Central Java, **Figure 1**.

East Java has the largest area among 6 provinces on the island of Java (about 407,803 km^2), and has the second largest population (40.67 million people). Based on province, West Java province has the largest population in Indonesia in 2020, which is 48.27 million people. Meanwhile, East Java province is bordered by the Java Sea in the north, the Bali Strait in the east, the Indian Ocean in the south, and Central Java Province in the west. Several small islands, namely the islands of Madura, Bawean, Kangean and a number of small islands in the Java Sea (Masalembu) and the Indian Ocean (Sempu Island and Nusa Barung) are also the East Java Region.

Administratively, the total number of Districts and Municipalities in East Java is 38, as seen in **Table 1**.

2. Disaster in East Java

Cumulatively, based on calculations from the records of the National Disaster Management Agency (BNPB), from 2013 to 2019, in East Java there have been 2676 hydrometeorological disasters. The details are: floods as many as 743 cases, landslides 514 cases, drought 66 cases, forest fires 361 cases, tidal waves 22 cases and strong winds as many as 970 cases. When viewed from a case-by-case approach, it can be seen that every year there is an increase in the number of disasters. Between 2013 and 2014 there were about 233 cases, then increased to 297 cases in 2015, increasing again in 2016 by 404 cases, 2017 around 434 cases, 2018 increasing to 455 cases and increasing rapidly in 2019 with the number of cases amounting to 620 cases. The trend of increasing hydrometeorological disasters in each type of disaster has undergone significant changes. This condition can be checked by conducting media studies, by looking for disasters that occurred in the period 2019 to 2020. Around 50 more local journalists wrote about hydrometeorological disasters, from floods, landslides, forest fires, droughts, tidal waves and droughts in East Java.

The East Java Regional Disaster Management Agency (BPBD) revealed that in 2019 there had been floods covering 15 districts, namely: Madiun, Nganjuk, Ngawi, Magetan, Sidoarjo, Kediri, Bojonegoro, Tuban, Probolinggo, Gresik, Pacitan,

No.	Region Name	Capital city	Area
1	District Bangkalan	Bangkalan	1,001.44 kms (2.10%)
2	District Banyuwangi	Banyuwangi	5,782.40 kms (12.10%)
3	District Blitar	Kanigoro	1,336.48 kms (2.80%)
4	Bojonegoro	Bojonegoro	2.198,79 kms (4.60%)
5	District Bondowoso	Bondowoso	1.525,97 kms (3.19%)
6	District Gresik	Gresik	1.191,25 kms (2.49%)
7	District Jember	Jember	3.092,34 kms (6.47%)
8	District Jombang	Jombang	1.115,09 kms (2.33%)
9	District Kediri	Ngasem	1.386,05 kms (2.90%)
10	District Lamongan	Lamongan	1.782,05 kms (3.73)
11	District Lumajang	District Lumajang	1.790,90 kms (3.75%)
12	District Madiun	Caruban	1.037,58 kms (2.17%)
13	District Magetan	Magetan	688.84 kms (1.44%)
14	District Malang	Kepanjen	3,530.65 kms (7.39%)
15	District Mojokerto	Mojosari	717.83 kms (1.50%)
16	District Nganjuk	Nganjuk	1,224.25 kms (2.56%)
17	Ngawi	Ngawi	1,295.98 kms (2.71%)
18	District Pacitan	Pacitan	1,389.92 kms (2.91%)
19	District Pamekasan	Pamekasan	792.24 kms (1.66%)
20	District Pasuruan	Bangil	1,474.02 kms (3.08%)
21	District Ponorogo	Ponorogo	1,305.70 kms (2.73%)
22	District Probolingge	Kraksaan	1,696.21 kms (3.55%)
23	District Sampang	Sampang	1,233.08 kms (2.58%)
24	Sidoarjo	Sidoarjo	634.38 kms (1.33%)
25	District Situbondo	Situbondo	1,669.87 kms (3.49%)
26	District Sumenep	Sumenep	1,998.54 kms (4.18%)
27	District Trenggalek	Trenggalek	1,147.22 kms (2.40%)
28	District Tuban	Tuban	1,834.15 kms (3.84%)
29	District Tulungagung	Tulungagung	1,055.65 kms (2.21%)
30	Municipality, Batu	Kota Batu	136.74 kms (0.29%)
31	Municipality, Blitar	Blitar	32.57 kms (0.07%)
32	Municipality, Kediri	Kediri	63.40 kms (0.13%)
33	Municipality, Madiun	Madiun	33.92 kms (0.07%)
34	Municipality, Malang	Malang	145.28 kms (0.30%)
35	Municipality, Mojokerto	Mojokerto	16.47 kms (0.03%)
36	Municipality, Pasuruan	Pasuruan	35.29 kms (0.07%)
37	Municipality, Probolinggo	Probolinggo	56.67 kms (0.12%)
38	Municipality, Surabaya	Surabaya	350.54 kms (0.73%)

Table 1.
The district and major cities in East Java.

Tranggalek, Ponorogo, Lamongan and Blitar. The worst flooding occurred in the Madiun area. The disaster affected 12,495 families from the population of 15 districts in East Java. In 2019, BPBD recorded areas affected by drought and lack of clean water covering 22 districts, 128 sub-districts and 450 villages. Meanwhile for forest and land fires, where East Java has 8 active volcanoes, forest fires occur when entering the peak of the dry season. The mountains are Arjuna, Welirang, Kawi, Kelud, Wilis, Semeru, Bromo, and Ijen. According to the Ministry of Environment and Forestry (KLHK) there were approximately 23,655 hectares of forest area and land burned in 2019.

In 2020, there are several records related to disasters such as strong winds that hit Ponorogo, landslides that occurred in Lumajang and floods that hit Pasuruan, Mojokerto, Jombang, Madiun and Trenggalek. In 2020, according to data from the Ministry of Environment and Forestry, there were around 19,148 hectares of land and forest burned in East Java. Throughout 2020, there were around 31 regions in East Java that experienced drought and lack of clean water. Meanwhile, in early 2021, East Java was faced with floods that hit several areas, the worst occurred in Jember which almost affected 7 sub-districts.

3. Field site study

This research was conducted in two places, namely Jawar Hamlet, Srimulyo Village, Malang Regency and Banaran Village, Ponorogo Regency, both of which are in East Java. These two areas have experienced land subsidence and landslides, as well as flash floods.

Jawar Hamlet, Sri Mulyo Village, Dampit Subdistrict, Malang Regency is one of the villages prone to landslides in East Java Province [3], Preliminary Study of Landslide In Sri Mulyo, Malang, Indonesia Using Resistivity Method And Drilling Core Data). On January 24, 2006, there was a landslide in the village which caused 1 house to be destroyed, 14 houses cracked on the walls and foundation. In addition, the landslide also resulted in 3 large landslide areas, and 12 small landslides that cut off access to village roads as deep as 60 cm. The area where the landslide occurred, after being examined geologically, was found to be located above weathered breccia rocks that lay on top of the limestone of the Wonosari Formation. When viewed from the angle of inclination, the landslide area actually has a slope of 12^0–20^0 or not too steep. The landslide incident showed that the cause of the landslide was not purely due to the slope, but because the area was located in a gravity fault area, which is thought to have caused the landslide of limestone, which is located at the bottom of the landslide area [4].

The village of Sri Mulyo, Dampit subdistrict (**Figure 2**) is geographically located at 8.2928 ^{0} SL 112.7991 ^{0} EL. Administratively, Sri Mulyo village in the north is bordered by Bumirejo Village, Ampelgading Sub-District in the east, Sumbermanjing Wetan Sub-District in the west, and Sukodono Village in the south. In general, the soil structure in Sri Mulyo village is podzolic soil, where the topography is plains and mountains with an altitude of 400–790 meters above sea level, and the slope of the slope is less than 40%. The average annual rainfall is 5229 millimeters. The majority of the population is coffee and salak farmers.

The second research area is Banaran village. Banaran village is located in Ponorogo Regency, East Java (**Figure 3**). Banaran village has a land area of 2827,713 ha divided into four small villages (hamlet), namely Krajan, Gondang Sari, Tangkil, and Sooro. Banaran Village is one of 18 villages in Pulung District. The locations of Banaran village is: In the west it is bordered by Bekiring Village, Pulung Sub-District; in the east bordering the Tambang Village, Pudak Sub-District; In the north it is bordered by Talun Village, Ngebel Sub-District and in the south by Wagir Kidul Village, Pulung Sub-District.

Figure 2.
Geoelectrical resistivity measurement survey design, Sri Mulyo Village.

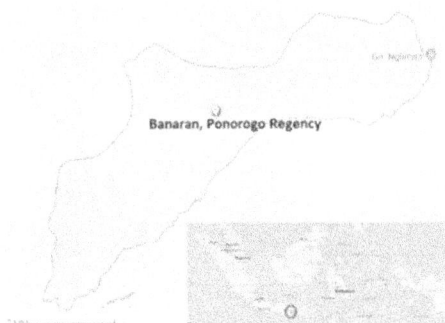

Figure 3.
Banaran Village, Ponorogo regency, Indonesia (Google maps, 2018).

One method to analyze a landslide is to use geophysical methods. Geophysical method is a method to determine the subsurface conditions of the earth based on physical parameters. The resistivity method is one of the most widely used geophysical methods in the fields of hydrogeology, disaster mitigation, and archeology [5–9]. Several previous studies has shown that the resistivity method is effective for knowing the subsurface conditions of landslide-prone areas [4, 10–14].

It is important to conduct research in the Jawar Hamlet, Sri Mulyo Village, Malang Regency and Banaran Village, Ponorogo Regency so that the area's vulnerability to landslides is known. This landslide analysis was carried out as one of the disaster mitigation efforts, because the area had experienced landslides in the past. The core drilling method was also carried out to confirm the results of the geoelectrical interpretation, especially for the Jawar hamlet area. The landslide analysis is expected to provide an overview of subsurface conditions supported by Turen and Ponorogo Geological Map, and it can determine the landslide fields as well thickness of landslide potential material at the study site.

4. Method

4.1 Sri Mulyo Village

The resistivity method basically utilizes the electrical properties of the earth, by interpreting the apparent resistivity parameters of subsurface rocks. This method is an active method, where an electric current is injected into the earth through two current electrodes (C1 and C2) then the resulting potential difference is captured by two potential electrodes (P1 and P2). By considering the geometry factor of the configuration used, it can be calculated the apparent resistivity of the rock below the surface.

In this study, the configuration used is dipole–dipole (**Figure 4**). Measurements are made by moving the potential electrode in a path with a fixed current electrode. Next, the current electrode is moved at a distance to the next "n", which is followed by moving the potential electrode along the cross-section. This is done until between C1 and P1 has a distance of "na", according to the conditions of the surface in the study area.

The research in Jawar Hamlet, Sri Mulyo Village was carried out in 2015 which is located at the coordinates of 08°18′44,86″ - 08011′05,16″ SL and 112049′22,02″-112041′56,47″ WL. There are 5 measurement lines for the resistivity method (**Figure 3**). Lines 1, 2, and 3 with a track length of 180 meters, 200 meters and 100 meters, respectively, and the spacing between electrodes is 10 meters. Lines 1 to 3 are measured from Southeast to Southwest. Line 4 with a track length of 200 meters with an electrode spacing of 10 meters while line 5 with a line length of 300 meters and an electrode spacing of 15 meters. Lines 4 and 5 are measured from West to East.

In this study, drilling points were also carried out. This drilling point is carried out in the red and yellow lines. This core extraction is used as a sample for laboratory tests in determining soil characteristics. Drilling was carried out at two points, namely point 1 located on the 60 meter track line 1 and point 2 located 70 meter line length line 2. The results of the drilling were used as supporting data for landslide analysis for the resistivity method. Geoelectric data processing is done with Re2dinv software. Interpretation is carried out by correlation with the Geological Map of the Turen Sheet.

4.2 Banaran Village, Ponorogo District

The method used in this research is geo-electric resistivity method, sama seperti yang dilakukan di desa Srimulyo. There are many configurations used in this method. In this research, we used a modified Wenner-Schlumberger configuration (**Figure 5**) with fixed electrode potential and the current electrode to obtain Vertical Electrical Sounding (VES). To obtain the lateral direction variation, the VES through interpolation dots are measured. This configuration is the right choice if the desired target is the VES with optimal field effectiveness and reduction accumulation error. Measurements were made at 12 points. The selection of this measurement point is based on the fact that the area is still relatively flat and the results of the interpretation can be correlated from one location to another.

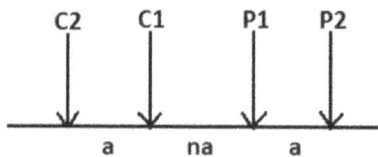

Figure 4.
Dipole–dipole configuration with, "a" spacing between electrodes (m), "ΔV" potential difference (mV), "I" injected current (mA), "n" number of layers, "ρ_a" apparent resistivity of rocks (ohm.m), and "k" geometry factor (m).

Figure 5.
Distribution of geo-electric resistivity measurement points.

5. Result and discussion

5.1 Jawar Hamlet, Srimulyo Village, Malang regency

Based on the results of resistivity data processing and information on the Turen Sheet Geological Map, it can be seen that Jawar Hamlet, Dampit Subdistrict is composed of three rock layers. The lithology of the local area is composed of clay (9,28–85,8 $\Omega.m$), tuff (178–779 $\Omega.m$) and breccia ($\geq$ 1629 $\Omega.m$). **Figure 6**, is 2D cross-sections of resistivity data processing, where the dotted line indicates the estimated landslide slip area at the study site, and **Figure 7** is 2D cross sectional. The slip plane in this study is the boundary between clay and tuff.

Figure 7 shows the presence of clay dominance on the three parallel lines. It is seen that clay predominates to a depth of about 10 meters below the ground surface. This will result in the weight of the soil during the rain will be even greater, due to the infiltration of rainwater into the soil which does not easily come out, due to the very low permeability of the clay. As a result, the boundary between clay and tuff will become increasingly slippery. If this continues for a long time, and if the slope is not able to withstand a large load, there will be a movement of soil down the slope which is commonly referred to as a landslide. The potential for landslides is also higher because the vegetation above the surface of the research site is in the form of seasonal plants (coffee), whose roots are not too deep. This type of plant (coffee), has roots that are not strong enough to bind soil grains.

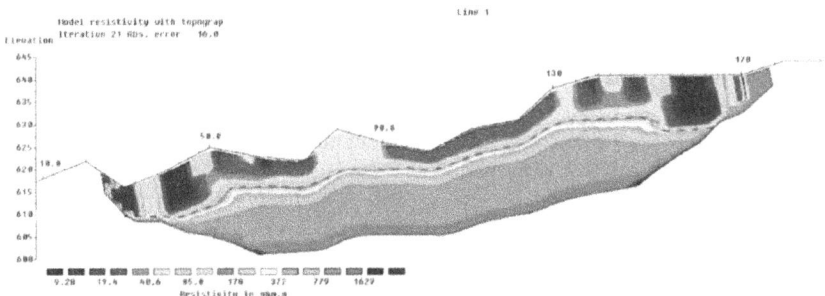

Figure 6.
2D cross section of resistivity line 1.

Figure 7.
Correlation of 2D cross-sectional resistivity lines 1, 2 and 3 respectively.

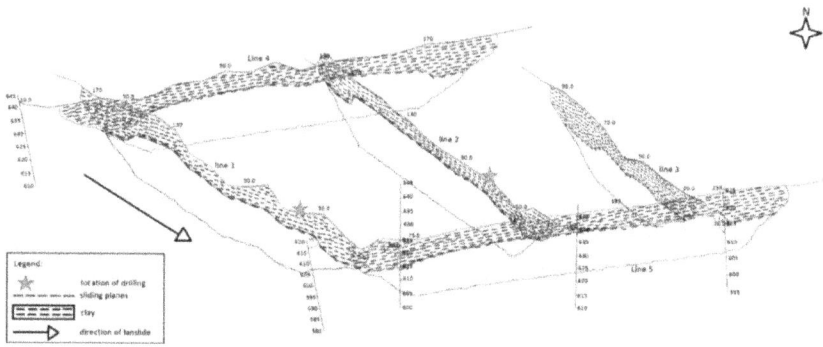

Figure 8.
Correlation of 2D resistivity and drilling points.

Figures 8 and **9** show the correlation between the resistivity data and the drilling data. Drilling was carried out on two measurement lines, namely line 1 and line 2 with a depth of 8 meters, which indicates that up to a depth of 8 meters the soil type at the study site was predominantly clay. This, in accordance with the interpretation of resistivity data from the five measurement lines, indicates that to a depth of about 10 meters, the study site is dominated by clay. The correlation of the five measurement lines shows that the avalanche direction is from Northwest to Southeast this is due to the difference in height between lines 4 and 5. When viewed from the results of the study, that the landslide area is relatively flat and the research location is relatively not steep, then the type of landslide in the research location is likely to be a creep type. This type of avalanche is a type that moves slowly down the slope.

Based on the correlation of all data, the data shows that the potential for landslides in the research location has a high level of vulnerability. This is because the thickness of the clay has exceeded 5 meters with a high average annual rainfall of 5299 mm/year, so this will increase the weight of the soil when it rains. In addition, the carrying capacity of plant vegetation is inadequate at the study site, causing settlements to be unsafe from landslide hazards. The results of this study may be one of the considerations of the local government in disaster mitigation at the research location so as to minimize casualties and losses due to landslides.

5.2 Banaran Village, Ponorogo District

The results of data processing show that the subsurface conditions at the research site are divided into four constituent rocks, namely; clay (0–100 Ω.m,) tuff

(100–1000 Ω.m), volcanic breccia (1000–3000 Ω.m), and andesitic lava (>3000 Ω.m). The depiction of the interpolation results for several depths can be seen in **Figure 10**. From the analysis results, the subsurface rock at the study site is dominated by clay. The further down, it was detected that the clay became more dominant. This indicates that the landslide material at the study site is very thick.

If the correlation is made for inline points, the following results (**Figure 11**) will be obtained:

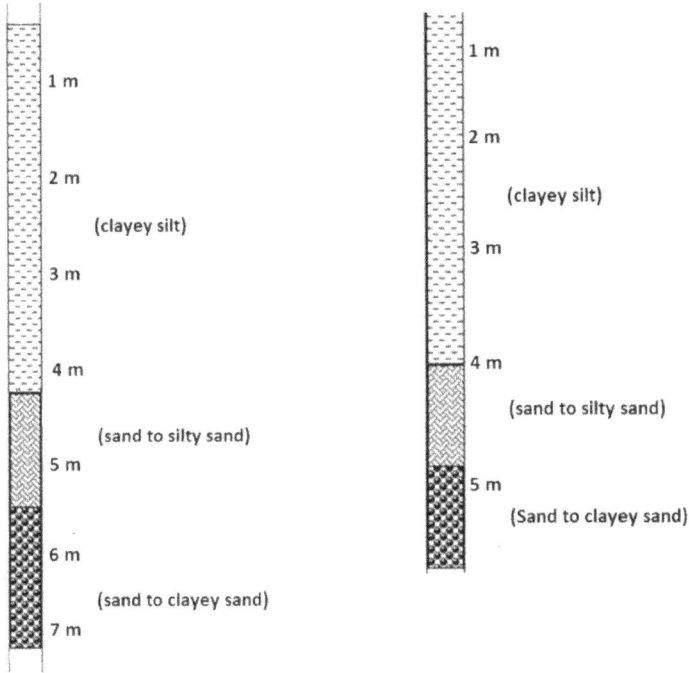

Figure 9.
Soil sample test results (a) line 1 and (b) line 2.

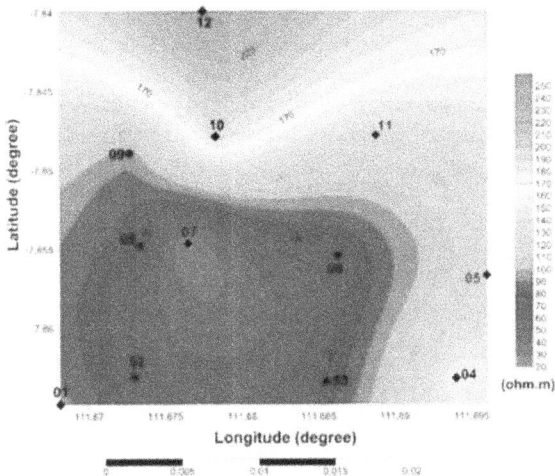

Figure 10.
Contour resistivity at 5 meters depth.

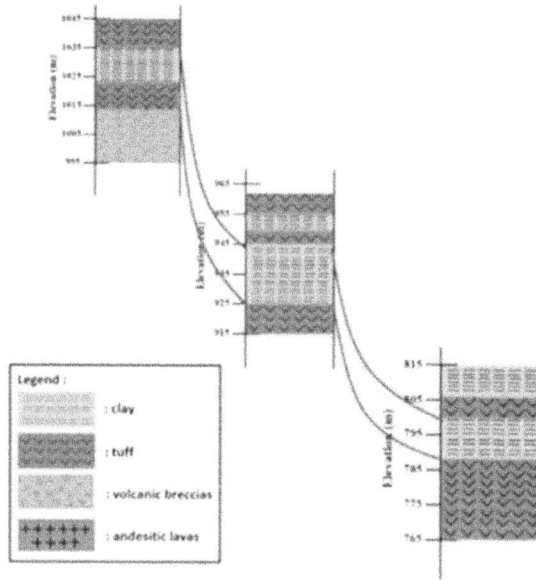

Figure 11.
Correlation of lithology of measuring points 11, 06, and 03 (from top to bottom).

The results of data processing from this resistivity method indicate that the landslide field is located at a depth of 8 to 35 meters. Tuff is indicated as a landslide field. Interpretation of resistivity data for lithological correlation shows that the upper part of the landslide area is dominated by rocks with low resistivity (conductive) and quite thick (5–35 m) which is indicated as clay. Clay is a type of rock that has very small permeability, which is 10^{-7} cm/s. In addition, the research location is a steep slope. From the interpretation of the lithological correlation, the estimated slope of the location is ≥ 400, and the rainfall was quite high at that time, namely 18.95 mm per day. The length of the rainy month is nine months.

The presence of clay with a thickness of more than 5 meters on steep slopes with high rainfall, will cause greater soil weight, especially when rainwater seeps into the soil. When it is found that the bottom layer is clay, the soil will be saturated with water and can no longer come out, because clay is almost impermeable. In addition, the soil which is dominated by clay will be very prone to landslides, because it is soft and slippery when it rains and cracks when it is hot. As a result, if it rains, the steep slope will accelerate the movement of the soil down the slope, which is known as a landslide. In addition, the use of land in the form of settlements and plantations of seasonal crops, such as ginger, spices and so on, makes plant roots not strong enough to withstand soil movement, and not strong enough to absorb rainwater. So, in the end there was a landslide.

6. Conclusion

The results of this study in Jawar Hamlet, Sri Mulyo Village indicate that the resistivity method can provide a good picture for investigations in locations that have the potential for landslides. The results of the correlation of resistivity data with drilling show that the research location is dominated by clay to a depth of about 10 meters. The level of vulnerability to landslides in Jawar Hamlet, Sri Mulyo

Village is high. The results of the investigation of the location's vulnerability to landslides are expected to be one of the references for the local government in efforts to mitigate landslides. One of the ways to do this is by replacing annual plants to strengthen the binding of the soil grains.

In addition, based on the interpretation of resistivity data, tuff and clay are indicated as slip planes, which are found at depths ranging from 8 meters to 35 meters. The landslide material is the part above the slip plane, which is dominated by clay with a thickness of 5–35 meters. Banaran Village, Ponorogo Regency is an area that is very prone to landslides, based on a landslide-prone area score. Parameters that actively support the occurrence of landslides at the study site are the presence of slip fields, steep slopes, high rainfall and inappropriate land use. Therefore, residents who live in the area are highly expected to be aware of landslides, which can occur at any time, especially during the rainy season. The level of vulnerability to landslides in the Banaran area is relatively high, so it is necessary to relocate residents' settlements around the Banaran area.

Acknowledgements

Thanks to Geophysics Laboratory, Physics Department, Brawijaya University in supporting the equipment. The research was funded by Brawijaya University.

Author details

Adi Susilo[1*], Sunaryo Sunaryo[1], Eko Andi Suryo[2], Turniningtyas Rachmawati[3] and Muwardi Sutasoma[1]

1 Geophysics Engineering, Brawijaya University, Malang, Indonesia

2 Civil Engineering, Brawijaya University, Malang, Indonesia

3 Urban Planning Engineering, Brawijaya University, Malang, Indonesia

*Address all correspondence to: adisusilo@ub.ac.id

IntechOpen

References

[1] Susanti., Pranatasari Dyah, Arina Miardini, Beny Harjadi, 2017. Analisis kerentanan tanah longsor sebagai dasar mitigasi di Kabupaten Banjarnegara (Vulnerability analysis as a basic for landslide mitigation in Banjarnegara Regency), Jurnal Penelitian Pengelolaan Lokasi Aliran Sungai (Journal of Watershed Management Research) Vol. 1 No. 1 April 2017 : 49-59.

[2] (https://commons.wikimedia.org/wiki/File:East_Java_Region_map.svg)

[3] Susilo, A., Sunaryo, Kosmas I., Rusli. Investigation of Jabung Temple Subsurface at Probolinggo, Indonesia using Resistivity and Geomagnetic Method. International Journal of Geomate. Vol. 13, 2017b, pp 74-80

[4] Wasis, Adi Susilo, dan Sunaryo, 2011. Perunutan Jalur Sesar Lokal di Desa Sri Mulyo Kecamatan Dampit, Kabupaten Malang berdasarkan data geofisika, Natural B, Jurnal Lingkungan dan Kesehatan, vol 1, no 1, pp. 41-50, April 2011.

[5] Susilo, A., Sunaryo, Kosmas I., Rusli. Investigation of Jabung Temple Subsurface at Probolinggo, Indonesia using Resistivity and Geomagnetic Method. International Journal of Geomate. Vol. 13, 2017, pp 74-80

[6] Susilo, A., Sunaryo, Alexander T. Sutanhaji, Fina Fitriah and M.F.R Hasan. 2017a. Identification of Underground River Flow in Karst Area Using Geoelectric and Self-Potential Methods in Druju Region, Southern Malang, Indonesia, International Journal of Applied Engineering Research, Volume 12, Number 12. pp. 10731-10738

[7] Susilo, A., Sunaryo, and Fina Fitriah. Fault Identification For Human Settlement in Pohgajih Village, Blitar, Indonesia Using Geoelectrical Resistivity Method for Hazard

Mitigation. International Journal of Geomate, Vol. 14, 2018, pp. 111-118

[8] Schmidt, A. Electrical and Magnetic Methods in Archaeological Prospection In S. *Campana* and S. Piero (eds) Seeing the Unseen. Geophysical and Landzchape Archaeology, 2009,pp 67-81.

[9] Giocoli, A. Cosimo, M., Paola, V., Sabatino, P., Enzo, R., Agata, S., Pierfransissco, B., Ciriaco, B., and Silvio, D.N. Electrical Resistivity Tomography Investigation in the Ufita Valley (Southern Italy). Annals of Geophysics. Vol 51, 2008, pp. 213-223

[10] Bell R., Jan-Erik K., Alejandro G., Thomas G., Bonn, Andreas H., and Braunschweig. 2006. Subsurface Investigation of Landslide using geophysical methods-geoelectrical appliclation in the Swabian Alb (Germany). *Geographica Helvetica*. 61. pp. 201-208

[11] Pilecki Z., 2017. Basic Principles for The Identification of Lanslide using geophysical methods. 3rd International Conferences on Applied Geophysics. 24. pp. 1-8. DOI: 10.1051/e3sconf/20172401001

[12] Jomard, H., T. Lebourg and E. Tric. Identification of the gravitational boundary in weathered gneiss by the geophysical survey: La Clapiere landslide (France). Journal of Applied Geophysics (Science Direct). Vol. 62, 2007. pp. 47-57.

[13] Perrone A, Sabatino P., and Vincenzo L. Electrical Resistivity Tomographies for Landslide Monitoring: A review. Berichte Geol., Vol. 93, 2017, pp. 129-134

[14] Kušnirák, D., Ivan D., Rene P., and Andrej M. Complex Geophysical Investigation of The Kapušany Landslide (Eastern Slovakia). De Gruyter, Vol. 16, 2016, pp. 111-124

Section 5

Application of
New Technologies

Chapter 10

Landslide Movement Monitoring with InSAR Technologies

Peifeng Ma, Yifei Cui, Weixi Wang, Hui Lin,
Yuanzhi Zhang and Yi Zheng

Abstract

Synthetic aperture radar interferometry (InSAR) is a technology that has been widely used in many areas, such as topographic mapping, land and resource survey, geological exploration, disaster prevention and mitigation, volcanic and seismic monitor and so on. Landslide, as a representative geohazard, include a wide range of phenomena involving downhill ground movement. InSAR, a technology which can measure surface deformation at the millimeter level over serveral days or years, is suitable to detect landslides with chronical and widespread movements. In this chapter, we introduce main process methods of InSAR data, including Persistent Scatter Interferometry (PSInSAR) and Distributed Scatter Interferometry (DSInSAR). A study area, Daguan County Town, one of the most landslide-prone areas in China is induced to demonstrate the practicability of InSAR in detecting landslides. Combined InSAR results with geological, geotechnical and meterological data, the distribution of landslide in Daguan County in spatial and temporal dimensions would be displayed. We also coupling numerical modeling and InSAR for characterizing landslide movements under multiple loads. The numerical results revealed that body loads dominated the cumulative downhill movements by squeezing water and air from voids, and precipitation caused seasonal movements with the direction perpendicular to the slope surface.

Keywords: InSAR, landslide movement, numerical modeling, spatial–temporal analysis, human activities

1. Introduction

Landslides represent a major geological hazard causing injury or death and causing economic loss [1–5]. More than 20,000 hazards associated with landslides occurred in China from 2013 to 2014, totally causing 10 billion CNY loss approximately [6]. Landslides tend to occurred in areas where natural or human activities are more frequent [7–11]. Exposing the causes of landslides is essential for characterizing slope mobility and hazard mitigation. The traditional method of landslide investigation is to conduct field surveys [10], but it is laborious and difficult to find the landslide boundaries. Synthetic aperture radar interferometry (InSAR) monitoring, based on satellite, is now an effective way for landslide deformation monitoring from regional to local scales [12–14]. In particular, the launch of Sentinel-1 satellites opens a new era of global coverage monitoring with revisit times of 12 days using one satellite and 6 days using two, making regular landslide monitoring feasible [15, 16]. Advanced multitemporal InSAR methods have been proposed to achieve centimeter-to-milli-meter deformation monitoring using multi-baseline SAR images by detecting tempo-rally coherent targets [17–19]. With advances in algorithm development, InSAR has been widely used to monitor slope movement [20–22], update landslide inventory maps [11, 23], and generate landslide risk maps [24, 25]. Multi-source space-air-ground monitoring technologies are also combined with InSAR to study landslide movements to overcome the limitations of single technology [11, 26].

Landslides in built-up areas should first be considered for monitoring, compar-ing with landslides in non-built-up areas. Human activities, e.g., land development, are frequent in these areas, so monitoring is necessary to prevent greater economic losses and casualities. In earthquake area zone of southeast China, with scarce land resources, the construction of new cities is often carried out on steep slopes. In these areas, infrastructures and buildings are at high risk since the land urbanization may destroy the original ecological environment [7]. InSAR can measurement surface deformation and the landslides boundaries can be confirmed through vertical deformation velocity. However, this method can only monitor the movement on surface. To reveal the physical mechanisms of landslide, the numerical modeling should be introduced to simulate physical process underground. Numerical model-ing is purely mathematical and thus is very different from physical modeling in the laboratory or field modeling [27]. The models are an abstract of the real geology object, so the geometries are usually simplified. Besides, the material properties of the simulated strata are retrieved through back analysis method based on InSAR measurements. The phenomena of landslides could be simulated after the definition of model, using Finite Element Analysis (FEA) method [28, 29].

The coupling analysis of numerical modeling and InSAR measurements have been proved an effective method to deduce the deformation of dams [30]. In this chapter, we review an experimental study [31] to show the performance of coupling numerical modeling and InSAR for monitoring slope. Daguan County Town, a built-up area located on a hillside, is considered as our study area. An advanced multi-temporal InSAR method would be used to measure surface movements with robust detection of persistent scatterers (PSs) and distributed scatterers (DSs). Then, the detection results would be validated by ground data from Global Naviga-tion Satellite System (GNSS) stations and inclinometer tubes. The coupling analysis of numerical modeling and InSAR measurement results would reveal the mecha-nisms of landslides, and influence of three main factors (e.g., precipitation, body loads and construction-induced loads) would be discussed.

2. Study area and datasets

Daguan County, located in the northeast of Yunnan and southwest of the Sich-uan earthquake fault zone (**Figure 1**), is one of the most landslide-prone areas in China [6]. From 1844 to 1974, nine earthquakes with magnitude greater than 5 occurred in Daguan County. Earthquake and fault activity have led to a wide distribution of ancient landslides in this area [32, 33]. The region has a subtropical monsoon climate, with the rainy season normally lasting from June to September [6]. The historical geological evolution of the county consists of three main periods of movements (i.e., the Jinning Movement, the Yanshan Movement and Jialing River Movement), which formed folds and fractures. The geological strata range from the oldest to the youngest include the Silurian (consisting of the Lower Huanggexi, Middle Shichuankan and Upper Caidianwan formations), the Devonian (consisting of the Lower Cuifengshan and Middle Qingmen formations) and the Quaternary (consisting of artificial, remnant slopes, fallen rocks and alluvial

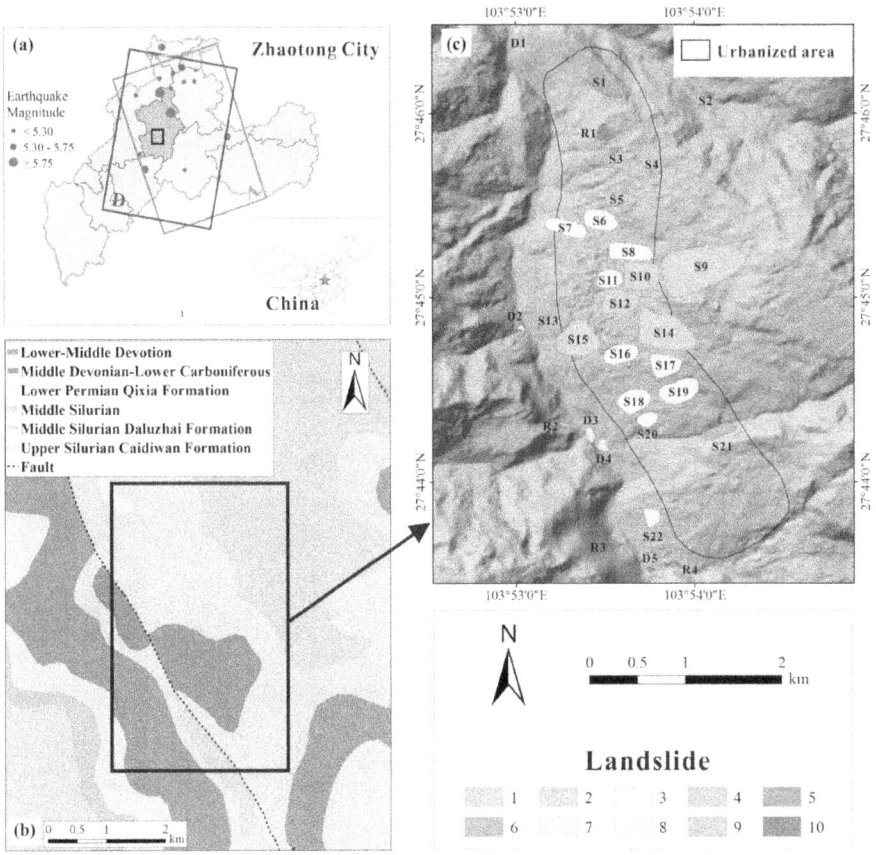

Figure 1.
(a) Study area and SAR images coverage. Rectangle a and D indicate the coverage of the ascending and descending Sentinel-1 images, respectively. The black rectangle indicates the location of Daguan. (b) the geological environment of Daguan County town. (c) Landslide inventory map. S1-S22 are the 22 landslides, D1-D5 are the 5 debris flows and R1-R4 indicate 4 rockfalls. The polygon colors indicate: (1) large and high risk; (2) large and medium risk; (3) medium and high risk; (4) medium and medium risk; (5) small and high risk; (6) small and medium risk landslides; (7) high risk and (8) medium risk debris flows; and (9) high and (10) medium risk rockfalls [31].

deposits) [33]. Accompanied by frequent crustal movements, the rocks in the area are extremely fragmented and therefore easily infiltrated by rain and groundwater. Prolonged rainfall can saturate, soften, and erode the soil, presumably leading to slope failure. The main soft and hard interaction surfaces that are prone to landslides include artificial and rockfall layers, joint fractures and relief joints, overlying soft soils and underlying bedrock, soft and hard rocks, and the interaction between different weathering surfaces. Totally speaking, the steep topography, tectonic action and stratigraphic lithology together shape the landslide-prone environment.

Daguan County Town is built on an ancient landslide (**Figure 2(a)**). The altitude ranges from $500m$ to $1500m$ and the angle of slope from 10° to 60°. Its main aspect of the slope is to the west. Human activity has significantly altered the landscape and topography, with buildings and infrastructure densely located in this area. In order to promote sustainable development, the Daguan County government began rezoning and revising the land use in 2010, and the land has been rapidly developed over the past few years. Construction works have triggered instability and response of ancient landslides by altering above-ground loads and subsurface geological and hydrological conditions [34, 35]. The Yunnan Geotechnical Engineering Survey and

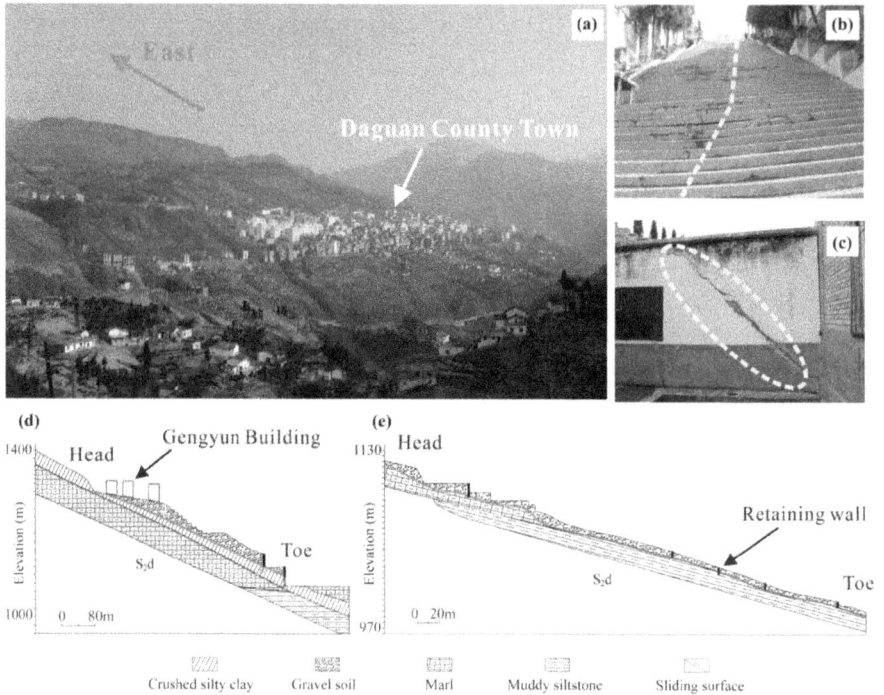

Figure 2.
(a) Photo of the Daguan County town. (b) Stair buckling in the region of S12. (c) Wall cracking in the region of S15. Stratigraphic cross-sections of S12 (d) and S15 (e) [31].

Design Institute conducted a field survey of the geological environment and distribution of geological hazards in the area in 2011 and generated an inventory map based on its identification of 31 landslides, including 22 landslides, 5 debris flows and 4 rockfalls (**Figure 1(c)**). **Figure 2(d)** and (**e**) are stratigraphic cross-sections of the two most severe landslides (e.g., S12 and S15). The geological setting consists of Silurian marl remants (S_2d) and artificial fill, and the sliding soils are loose Quaternary gravelly soils. The slides of S12 and S15 caused buckling of the stairs and cracking of the walls. Daguan No.1 Middle School is located close to the top of the landslide in S12 (**Figure 2(d)**), and a building (e.g., Genyun Building) here was rebuilt in May 2018. In S15, near the toe of the slope, is the retaining wall (**Figure 2(e)**), whose construction began in June 2018, to dampen movement by applying lateral foces. These construction works may affect the mobility of the landslide.

The SAR images is from Sentinel-1 in the wide swath imaging mode. The monitoring coverage is as shown in **Figure 1(a)**. A total of 60 ascending images from February 12, 2017 to February 26, 2019 and 56 descending images from February 19, 2017 to March 5, 2019.

3. Methods

3.1 Deformation estimation of PS and DS points in a two-tier network

3.1.1 Detection of the most reliable PS in the first-tier network

The problem of decorrelation usually occur in landslide areas. To solve this problem, we used a two-tier network to detect PSs and DSs with robust deformation estimation [36–38], which can help improve the measurement points density. The first-tier network was constructed to find the most reliable PSs over the study area. Depend on amplitude dispersion, the primary PSs candidates (PPSC) could be selected firstly [39]. A Delaunay Triangulation Network (DTN) could then be constructed using PPSC. Some additional redundant arcs should be added to improve the density of the DTN, because of the sparse PPSC [37]. After removing the atmospheric phase screen (APS), the signal model of N observations was written like [40, 41]:

$$y = A\gamma \tag{1}$$

where $y = [y_1, \dots, y_N]^T$ ($(\cdot)^T$ is the transpose operation) represents the complex values of interferograms after removing the APS, γ is the reflectivity vector and A is the sensing matrix containing the steering vector $a(\Delta h, \Delta v)$ as its columns:

$$a(\Delta h, \Delta v) = [\exp(j2\pi(\xi_1\Delta h + \eta_1\Delta v)), \dots, \exp(j2\pi(\xi_N\Delta h + \eta_N\Delta v))]^T \tag{2}$$

where $\xi_i = 2b_1/(\lambda r_0 \sin\beta)$ (b_i is the perpendicular baseline, λ is the wavelength, r_0 is the slant range and β is the incidence angle) and $\eta_i = 2t_i/\lambda$ (t_i is the temporal baseline). Δh and Δv are relative height and mean deformation velocity to be determined, respectively. We used beamforming and an M-estimator to calculate the relative estimates [36]. The beamforming-based inversion is as follows:

$$\gamma(\Delta h, \Delta v) = \frac{\left|a(\Delta h, \Delta v)^H y\right|}{\|a(\Delta h, \Delta v)\|_2 \|y\|_2} \tag{3}$$

where $|\cdot|$ is modulus operation for each element, $\|\cdot\|_2$ is 2-norm, and $(\cdot)^H$ stands for the conjugate-transpose operation. The maximum value of γ is calculated by sampling Δh and Δv to describe the temporal coherence of points. The arc between points would retain if the corresponding temporal coherence is larger than a given threshold; otherwise, the arc would be removed. The threshold we set here is 0.72, typical for millimeter-level deformation estimation [42]. For the preserved arcs, the preliminary estimates were used to unwrap the temporal phase. Then the inversion problem is transformed to:

$$\Delta\Phi = DJ = \begin{bmatrix} 2\pi\xi_1 & 2\pi\eta_1 \\ \vdots & \vdots \\ 2\pi\xi_N & 2\pi\eta_N \end{bmatrix} \begin{bmatrix} \Delta h \\ \Delta v \end{bmatrix} \tag{4}$$

where $\Delta\Phi$ is the unwrapped phase. In the presence of low-quality images, the preliminary relative estimates may be biased. To address this issue, we introduced an M-estimator to re-calculate the estimates using the unwrapped phase [43]:

$$J = (D^T W_M D)^{-1} D^T W_M \Delta\Phi \tag{5}$$

where W_M is an iteratively computed weight matrix that represents the phase quality. Compared to the unweighted least square estimate, the M-estimate assigns smaller weights to the phase outliers, thus improving the robustness of the estimate. After solving the relative estimates, we integrate them with the network adjustment method. When the adjustment matrix is poorly conditioned, its inversion may be unstable. To solve this problem, we apply a ridge estimator to the network adjustment [44]. It introduces a conditioning matrix σI (I is the identity matrix) in the inversion:

$$X = \left(G^{\mathrm{T}} W_R + \sigma I\right)^{-1} G^{\mathrm{T}} W_R H \qquad (6)$$

where X contains the absolute estimates of PS, G is the conditioning matrix consisting of $-1, 0$ and 1 (-1 and 1 represent the start and end of the arc, respectively), W_R is a diagonally weighted matrix containing the temporal coherence, and H contains the arc on relative estimates. The adjustment parameter σ is determined according to the L-curve method [44]. The ridge estimator outperforms the traditional least-square estimator in regulating possible ill-conditioned problems. The effectiveness of the M-estimator and the ridge estimator has been evaluated in [36] and is omitted here for simplicity.

3.1.2 Detection of the remaining PS and all the PS in the second-tier network

The PSs detected in the first-tier network were regarded as reference points to build the second-tier network. Identifying statistically homogeneous pixels (SHPs) by Kolmogorov–Smirnov (KS), the complex covariance matrix (CCM) C could be calculated. Then the inversion of C could be used to change the optical phase by the Broyden–Fletcher–Goldfarb–Shanno (BFGS) algorithm. It should be noted that C is rank deficient, and the inversion is not stable when the amount of SHPs is less than N. To solve this problem, we have revised it as following [19, 45]:

$$\theta = \arg\max_{\theta} \left(\Lambda^{\mathrm{H}}(C \circ \Psi)\Lambda\right) = \arg\max_{\theta} \left(\Lambda^{\mathrm{H}} C \Lambda\right) \qquad (7)$$

This expression improves the robustness of estimation by assigning a larger weight to the higher-coherence phase and avoiding matrix inversion. Then, we employed a more efficient phase-linking method to obtain the optimal phase [46]. This method is called a coherence-weighted phase-linking (CWPL) method [45]. Finally, we used the reconstructed optimal phase to identify whether the DSC is a true DS using temporal coherence (T_DS) thresholding (0.65 in this case). The workflow we can see on the top left of **Figure 3**.

3.2 2D deformation velocity decomposition

Landslides are always described as a movement with a predominantly vertical orientation. However, a single track only provides deformation in the LOS direction. To describe the movement of landslides properly, we appreciated that the up-down and east–west components of the deformation could be calculated using the observations obtained from both ascending and descending orbits [47]. Suppose that in the Cartesian coordinate system, the direction of the X-axis is east, the direction of the Y-axis is north, and the Z-axis is up. The deformation of a target on the earth's surface is:

$$U = U_x s_x + U_y s_y + U_z s_z \qquad (8)$$

where U_x, U_y and U_z are the eastern, northern, and vertical components of U (the real movement of a landslide), and s_x, s_y, and s_z are unite vectors in the respective directions [48]. Because of the polar-orbit of Sentinel-1, the LOS deformation is insensitive to movement in the north–south direction, and the U_y is negligible. U_A and U_D were used to represent the LOS deformation velocity for the ascending and descending, respectively. It should be noted that we assumed that the

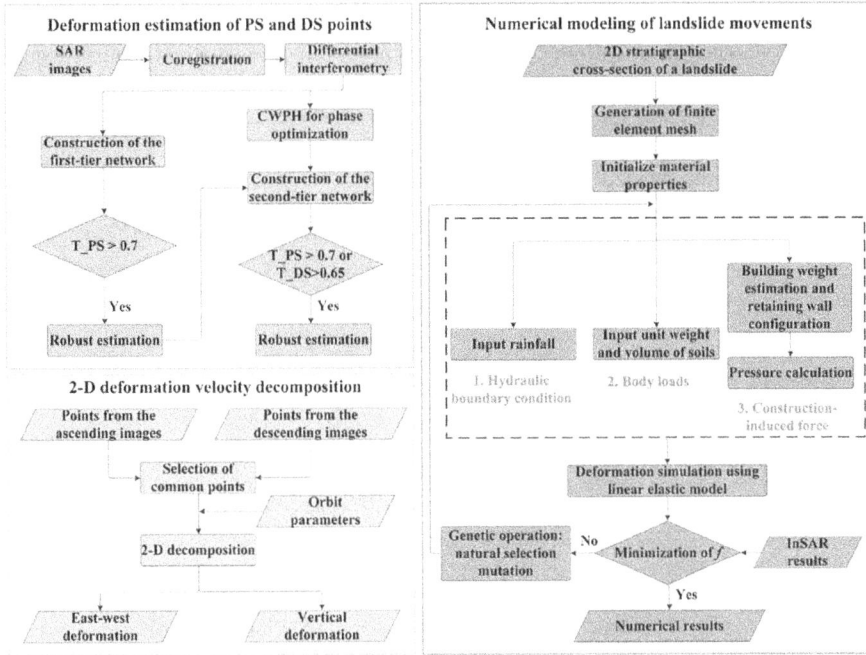

Figure 3.
Workflow of PS and DS detection, 2D deformation decomposition, and numerical modeling [31].

mean velocity of the landslide motion is the same for both types of orbit images, since their time spans are similar. Therefore, the projections of the east–west and up-down motions in the LOS direction could be written as:

$$\begin{cases} U_a \approx U_x a_x + U_z a_z \\ U_d \approx U_x a_x + U_z a_z \end{cases} \tag{9}$$

where a_x and a_z represent the unit LOS vectors obtained from the orbit parameters (**Figure 3**).

3.3 Numerical modeling of landslide movements using finite element analysis

In order to describe the underground process of landslide movement, we performed numerical model of landslide movement in GeoStudio software [49, 50]. SIGMA/W is a tool in GeoStudio that can perform stress and deformation analyses of earth structures, we can use it to simulate the physical process of ground volume change in response to self or external loading. Before solving, three components should be identified, which are geometry, material properties and boundary conditions. The geometry is the cross-section of a slope, it can be defined by simplifying the stratigraphic data.

The settings of material properties is crucial to make the model we defined close to the reality. We decided the material properties by combining the empirical knowledge and back analysis method [30, 51]. The purpose of Back analysis is to make the deformation of numerical modeling and InSAR measurement results consistent by modifying material properties. We first defined an objective function using the 2-norm of the difference between the modeled and measured deformation:

$$f\left(p_1, p_2, \ldots, p_m\right) = \sqrt{\frac{1}{n} \sum_{i=1}^{n} U_i^* - U_i} \qquad (10)$$

where $\left(p_1, p_2, \ldots, p_m\right)$ is the material properties set to be defined, U_i^* and U_i are simulated and InSAR measurement results, respectively. $|\cdot|$ is the absolute value operation and n means the number of InSAR scatterers located at the landslide. We then used the genetic algorithm (GA) to iteratively search for the optimal material properties [52]. The range and precision of soil properties were initialized in the GA based on a empirical knowledge [53–55]. The optimal set of property is the combination of material parameters by minimizing the objective function. If the objective function was not minimized, the GA is subjected to natural selection and mutation with crossover and mutation probability set to 0.9 and 0.1, respectively. Considering the case of non-convergence, we set an upper limit on the number of iterations to 50, which is also considered optimal when it is reached.

The boundary conditions were used to replicate real-world loading in deformation simulation. Considering the actual forces of the study area, three kinds of boundary condition are introduced in modeling. The first one is body load, which represents the body gravity related to the volume of the element and material, and it can be calculated by the unit weight and volume of soils. The second one is precipitation-induced hydraulic condition. The precipitation-induced hydraulic condition was regarded as a function varied with time. The input precipitation data was collected during the acquisition time of Sentinel-1 images. Amplitude of InSAR seasonal deformation varies with altitude under the influence of precipitation accumulation and stratified atmospheric delay [56, 57]. To calibrate this effect, we set different rainfall intensity at different elevations. The third boundary condition is construction-induced force (e.g., building loads). The building pressure was estimated by the concrete-steel density, thickness of walls and number of floors in the buildings obtained from field survey. After setting boundary conditions, we used the linear elastic model to simulate the deformation [58]. Linear elastic model assumes that stress is proportional to strain, and the load–displacement response is likely linear elastic along the lower initial portion of the stress–strain curve (**Figure 3**).

4. Experimental results

4.1 InSAR results and validation

4.1.1 2-D deformation mapping

Before decomposition of deformation, we identified the common points detected from ascending and descending tracks. The InSAR results are converted to raster data at $20m \times 20m$ resolution, the value of each grid cell was taken as the mean value of all points within it, and the position was in the center of each grid cell. **Figure 4(a)** and **Figure 4(b)** present the east–west and up-down direction deformation velocity map, respectively. The positive and negative velocities in **Figure 4(a)** indicate the eastward and westward motion, respectively, whereas the positive and negative velocities in **Figure 4(b)** indicate vertical uplift and subsidence, respectively. Sparse points were observed in the area with low elevation close to the Daguan River, caused mainly by slope-induced shadow and layover problems. High-elevation area also displayed a sparse distribution of points, caused

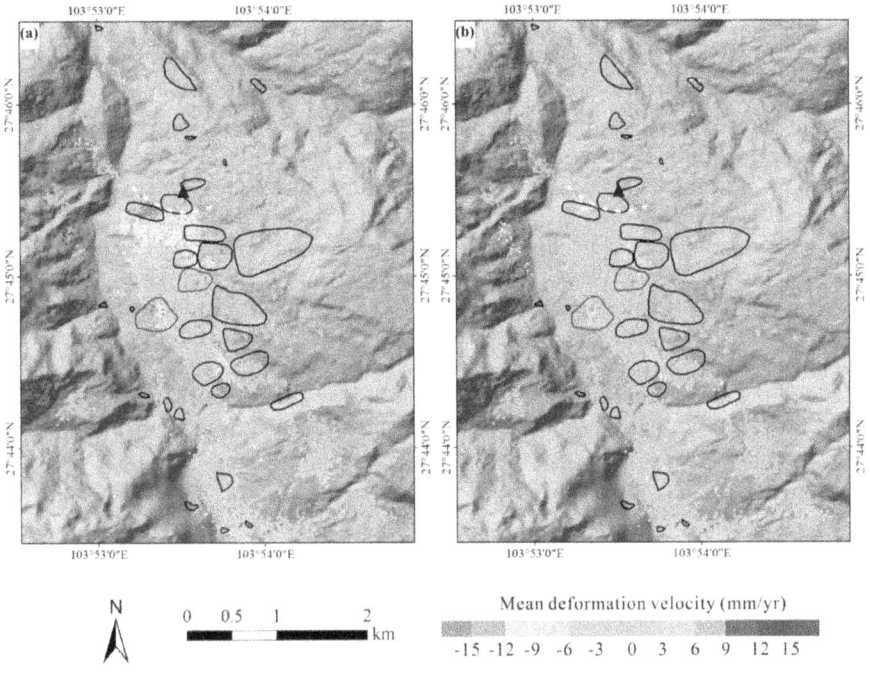

Figure 4.
Spatial pattern of InSAR measurements. (a) and (b) deformation velocity in east–west and vertical direction, respectively. The black triangles indicate the locations of the reference point and the white dashed line marks the boundary of the moving and stable areas. The polygons in the map indicate the landslide boundaries from the inventory map, and two landslides with blue boundaries are S12 and S15 [31].

mainly by vegetation-induced decorrelation effects. The results demonstrated that the moving direction was generally westward and downward, following the downhill direction. The westward movement was more significant than the downward movement because the slope angles of most positions were less than 45° and the horizontal projection of downhill movements was larger than the vertical projection. The overlap between the landslide polygons and 2D deformation suggests that the movements generally agreed well with the landslide areas. However, the spatial extents of landslides and monitored movements differed, indicating a possible change in landslide mobility from the time of the field survey (2011) to the acquisition time of the Sentinel-1 images (2017–2019). There was a sharp boundary between the moving and stable areas in the north of the study area (**Figure 4**). We, therefore, infer a fault here. Considering that Fenping Fault crossed the study area in **Figure 1(b)**, we conclude that the sharp boundary indicates the location of Fenping Fault. The Fault acts as barriers to groundwater flow, resulting in large differential deformation between the two sides of the fault. Considering the types of landslide may influence the interpretation of measurement results and subsequent numerical modeling, we confirmed that all landslides in this area are translational slides from field surveys. The polygon with more than 10 InSAR points within the boundary would be considered to be detected successfully. So, 24 of 31 landslides in the inventory map were successfully detected. The undetected landslides are almost small-sized (smaller than $0.01km^2$) except S21. There are some landslides without sufficient points or without significant movements (larger than $5mm/yr$), but we cannot guarantee that they are stable because rapidly moving targets may be shadowed or exceed the detection limitation of InSAR methods [59].

4.1.2 Validation of InSAR measurements using ground data

As the yellow labels shown in **Figure 5**, there are three global navigation satellite system (GNSS) stations (G1, G2 and G3) and two inclinometer tubes (I1 and I2) in the study area. We selected neighboring points from InSAR measurement results and ground monitor stations for comparison.

The measurements from GNSS are in three directions. We selected GNSS measurements along east–west and up-down directions to compare with InSAR 2-D deformation results. The linear deformation velocity was derived by linear fitting function. It should be noticed that there are missing data at G2 and G3, so we extend the time span of fitted data to July 2019 for them. The results implicated that the movements in horizontal were more significant than that in vertical movements. We can see from the mean deformation velocity that most InSAR results agreed well with the GNSS results except the horizontal movement of G2. That may be due to the fact that the InSAR point selected for comparison was located on a relatively stable structure rather than moving ground.

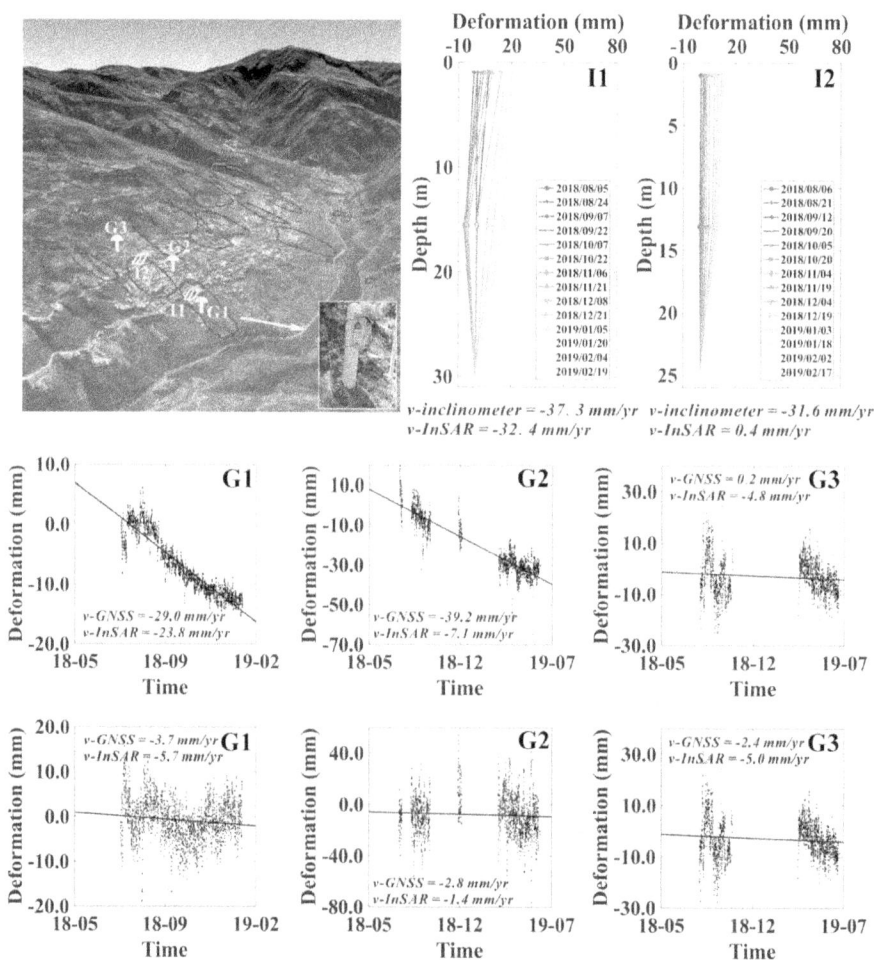

Figure 5.
Comparison between InSAR and ground data. (left top) location of GNSS stations and inclinometer tubes. (right top) comparison between InSAR and inclinometer data. (bottom) comparison between InSAR and GNSS data, horizontal movement in the upper row and vertical movement in the lower row [31].

Inclinometer tubes were installed to measure the deformation at different depths. The oscillation range of these two tubes was ±15°. We can obtain the deformation of I1 at depth of 1*m*, 15*m* and 30*m*, the deformation of I2 at depth of 1*m*, 13*m* and 25*m*. The measurement result provided two direction movement, x-direction is the tangential direction of the ground, y-direction is the vertical direction. We assumed that x-direction movement is along western because it is the same as the slope orientation, and the x-direction movement of I1 and I2 at different depth from August 2018 to February 2019 are shown on **Figure 5** (Right top). We use the deformation velocity at the depth of 1*m* to validate InSAR measurement results. The InSAR results agreed well with I1, but not with I2. The difference is 4.9*mm*/*yr* and 31.2*mm*/*yr*, respectively. The reason is that the InSAR scatterer selected for comparison may be located on a relatively stable structure. After validation, we can reasonably integrate them into numerical modeling for geological parameter retrieval.

4.2 Numerical modeling results

Two cross-sections of slopes (S12 and S15) are selected for numerical modeling analysis since they are typical slopes in this area and had been mapped by field survey. The size of each element of models was set to 15*m*. The InSAR measurement results of S12 and S15 were coupled to estimate the optimal material properties using GA. **Table 1** shows the range, precision and the optimal result of eight main parameters we used. The external stress loading was set as a function with four variables, and they are Max depth, K-Modulus, N-exponent and k(0). Cohesion and friction angle were set according to empirical knowledge. The minimized deformation is 1.3*mm*/*yr* and the corresponding material parameters were assumed to be optimal.

4.2.1 Numerical modeling of landslide S12

We combined the 2-D deformation to calculate the downhill movements of S12 (**Figure 6(a)**). The maximum combined movement was 23.2*mm*/*yr* and the direction was generally consistent with the downhill direction. Numerical modeling was conducted to derive the movements of S12. By iteratively searching for the optimal soil properties using GA, surface deformation by numerical modeling became consistent with measured movements by InSAR, and then we assumed that the material property set was valid. The cumulative deformation by numerical modeling is shown in **Figure 6(c)**. InSAR can measure only surface deformation, whereas numerical modeling depicted full-scale movements of S12 from the surface to the bottom. Deformation at the surface was generally more significant than that at the bottom, consistent with inclinometer data. Daguan No. 1 Middle School is close to the head of the slope, and the weights of three main buildings (Science and

Material property	E-Modulus Max depth (m)	K-Modulus	N-exponent	k(0)	Poisson's ratio	Pore water pressure (KPa)	Cohesion (MPa)	Friction angle (degree)
Range	[11, 20]	[160, 250]	[0.1, 0.5]	[1.1, 2]	[0.1, 0.4]	[−600, 0]	0.2	20
Precision	1	10	0.1	0.1	0.05	100	0	0
Result	15	200	0.2	1.2	0.2	−200	0.2	20

Table 1.
Main material properties estimated by the GA.

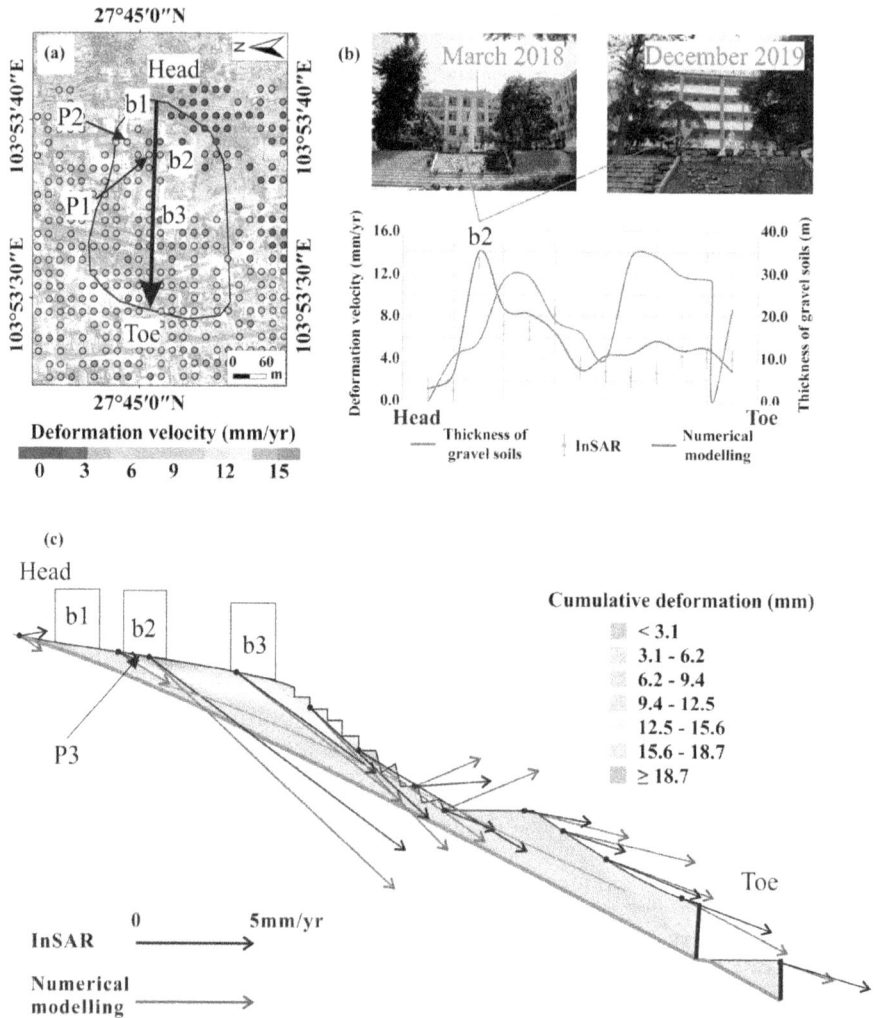

Figure 6.
*(a) Combined deformation velocity of the landslide S12 by InSAR. The background is optical image from Google earth. P1 and P2 are selected for time series analysis in **figure 7**. (b) Relationship between the thickness of gravel soils and deformation by InSAR and numerical modeling. Two photos show the Gengyun building before and after reconstruction, respectively. (c) Simulated deformation of the sliding layer in numerical modeling. The black and purple arrows indicate the surface downhill movements by InSAR and numerical modeling, respectively. The rectangles b1, b2 and b3 indicate the locations of science and technology building, Gengyun building, and Zizhi building, respectively. The black solid rectangles indicate the retaining walls. P3 is selected for time series analysis in **figure 7**.*

Technology building, Gengyun Building, and Zizhi Building) along the cross-section were estimated to impose building-induced loads, aggravating landslide movements. In particular, Genyun Building was rebuilt during this time, and led to the maximum cumulative deformation of 34.9mm in numerical modeling. In contrast, there are no buildings or infrastructures at the toe of S12, and the retaining walls mitigate downhill movements by imposing lateral loading. That makes the movement here less than that of the head, despite the relatively equal thickness of the soil layers (i.e., similar body load). In the central position of S12, the gravel soil is thin due to the location of Cuiping Road, which reduces the body load, and the movement here is relatively small.

Two moving scatterers (P1 and P2 in **Figure 6(a)**) are selected to study the temporal evolution (**Figure 7**). The independent deformation induced by three boundary conditions is also demonstrated. In the ascending track of Sentinel-1, seasonal movement is highly correlated with precipitation with a 1 to 3 month delay. The boundary conditions we defined for Daguan County are body loads, precipitation-induced hydraulic changes and construction-induced forces. Since the model has shown to be close to the measurements, we control for a single boundary condition to discuss the effect of each boundary condition on the landslide and the results are shown in **Figure 7**. The simulated time series deformation under precipitation-induced hydraulic condition showed similar trend with InSAR results (**Figure 7(a)** and **(c)**). We conclude that the seasonal trend is caused by precipitation, which drives movements by changing pore pressures that decrease when surface water cannot infiltrate the landslide body, and increase when surface water infiltrates the landslide body [60–62]. The delay is related to the pore pressure diffusion time since the onset of intense precipitation [63]. Interestingly, seasonal movement was not distinct from the descending track. This is because the LOS direction of the descending track is generally parallel to the slope surface. The simulated results showed that the direction of seasonal deformation was generally perpendicular to the surface (**Figure 7(f)**). Consequently, the LOS deformation from the descending track is insensitive to the seasonal rebound and subsidence. In the descending track, P1 and P2 showed continuous movements away from the sensor, which indicated a continuous downhill movement. Gravel soils were consolidated in response to body loads due to squeezing of water and air from the voids. The resulted movements were continuous and showed a decelerating trend with the increased consolidation (**Figure 7(d)**). Body loads caused larger cumulative deformation than the other two boundary conditions, indicating that it dominates

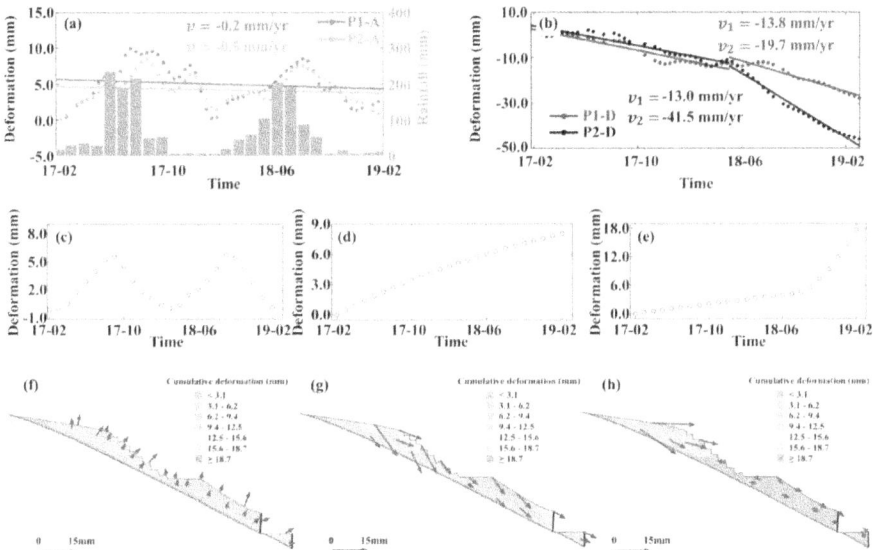

Figure 7.
Time series deformation of P1 and P2 in the (a) ascending (P1-a and P2-a) and (b) descending (P1-D and P2-D) images. Linear fitting is conducted before and after may 2018 for P1-D and P2-D. monthly rainfall data is collected from the National Meteorological Information Center. (c–e) Are time series deformation of P3 numerically modeled by precipitation-induced hydraulic change, body loads, and construction-induced force, respectively. (f–h) Are cumulative deformation numerically modeled by precipitation-induced hydraulic change, body loads, and construction-induced force, respectively [31].

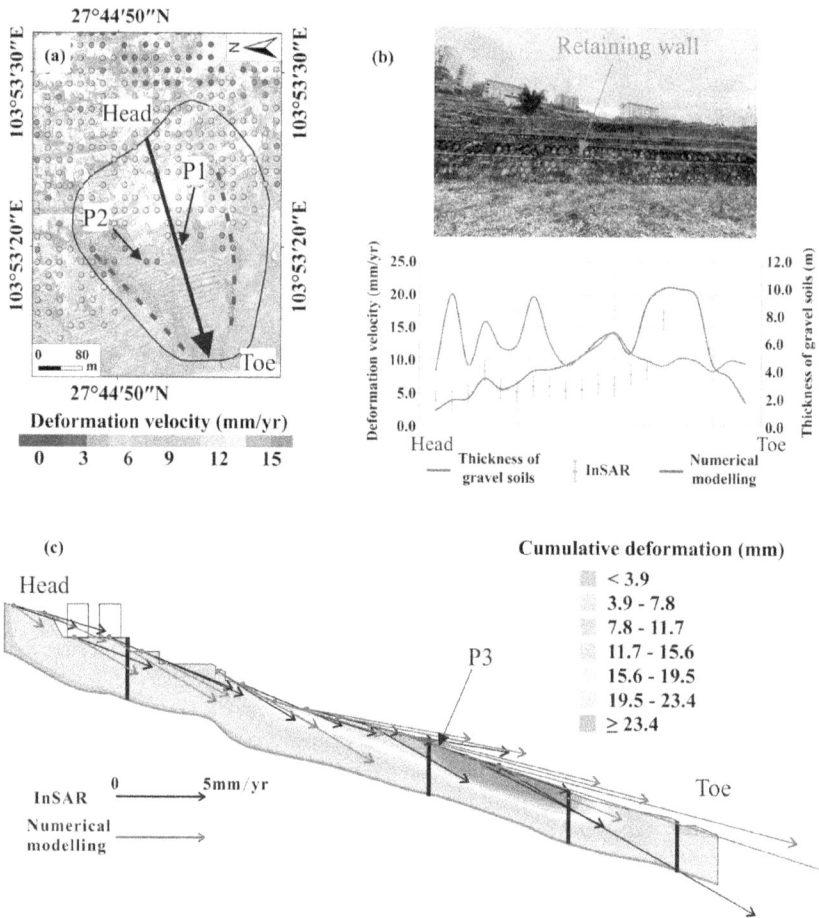

Figure 8.
(a) Combined deformation velocity of the landslide S15 by InSAR. The background is optical image from Google earth. P1 and P2 are selected for time series analysis in figure 9. The blue lines indicate the gullies. (b) Relationship between the thickness of gravel soils and deformation by InSAR and numerical modeling. The photo shows the retaining walls. (c) Simulated deformation of the sliding layer in numerical modeling. The thickness of cross-section has been increased by 4 times for better visualization. The black and purple arrows indicate the surface downhill movements by InSAR and numerical modeling, respectively. The black solid rectangles indicate the retaining walls. P3 is selected for time series analysis in figure 9.

cumulative downhill movements. The continuous movements of P1-D and P2-D implicated a constant velocity before May 2018 and a subsequent acceleration, suggesting different causes. As described above, the Gengyun Building started reconstruction in May 2018. Construction works caused additional loading associated with softened soils and hydrological change [64]. To model it, we defined that Gengyun Building-induced force started to be effective from May 2018 and the pressure increase from 0 to 48kPa gradually based on the calculated building weight. In this sense, Gengyun Building-induced force became a transient boundary condition during the monitoring period. The simulated results showed that P3 was relatively stable at the beginning and moved significantly after May 2018 due to reconstruction works of Gengyun Building (**Figure 7(e)**). This component was added to the movements caused by body loads and yielded an acceleration in time series deformation. Compared with seasonal fluctuation caused by precipitation, construction works caused permanent change of the time series trend.

4.2.2 Numerical modeling of landslide S15

Landslide S15 is located lower than S12 and has caused wall cracking and fracturing at Daguan Vocational School (**Figure 2(c)**). InSAR points were only present in the upper part of S15 because the lower part is covered by vegetation (**Figure 8(a)**). The maximum combined movement velocity was 21.4*mm/yr*. Two buildings were located close to the landslide head. However, because they were constructed many years ago, gravel soils have been consolidated and measured movements were not significant at the head. The relationship between the thickness of gravel soils and deformation implicated that the movements were more significant in the lower part. The effective modulus of elasticity in the lower part of S15 calculated by GA was smaller than that in the upper part. Two gullies formed the landslide boundary (**Figure 8(c)**). In rainy seasons, the rainfall converged to the gullies, ingressing and washing away the soils and decreasing slope stiffness [65, 66]. That decreased the effective modulus of elasticity of the sliding layer. The movements were therefore more significant in the lower part of S15. The largest cumulative deformation is 54.7*mm*. The movements became small at the landslide toe, because retaining walls were used to maintain stability of Zhaoma Highway therein.

Similar to S12, the seasonal movements were significant in the ascending Sentinel-1 images and were less distinct in the descending images (**Figure 9**). The seasonal trend of S15 showed a rebound in winter and subsidence in summer, which is opposite to the seasonal variance of S12 and precipitation. That may be caused by the different positions of the reference points when processing the data. The phenomenon that amplitude of seasonal trend is different at different elevations has been studied in [56, 57]. Hu et al. [57] attributes it to different accumulated precipitation at different elevations. The total precipitation accumulated during the wet seasons in the mountains at higher elevation is larger than that in the valleys/basins at lower

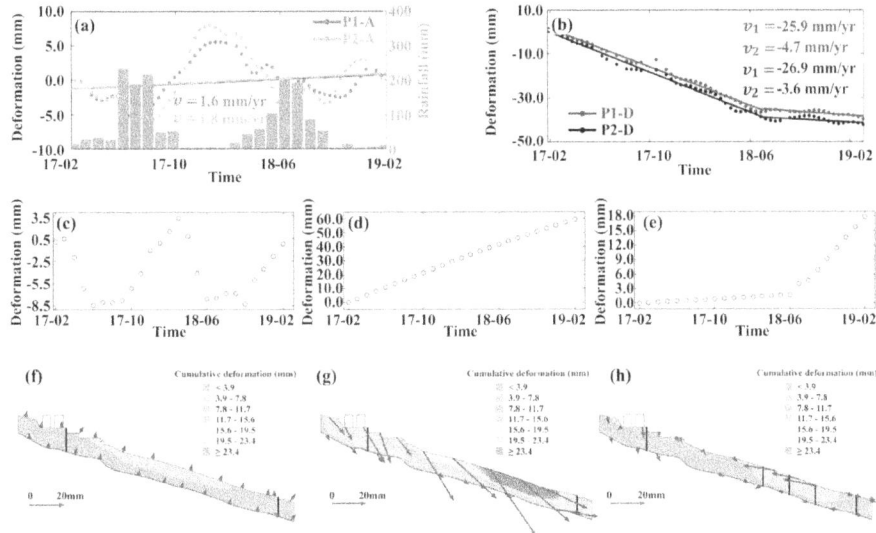

Figure 9.
Time series deformation of P1 and P2 in the (a) ascending (P1-a and P2-a) and (b) descending (P1-D and P2-D) images. Linear fitting is conducted before and after July 2018 for P1-D and P2-D. monthly rainfall data is collected from the National Meteorological Information Center. (c–e) Are time series deformation of P3 numerically modeled by precipitation-induced hydraulic change, body loads, and construction-induced force, respectively. (f–h) Are cumulative deformation numerically modeled by precipitation-induced hydraulic change, body loads, and construction-induced force, respectively. The thickness of cross-section has been increased by 4 times for better visualization [31].

elevation. Dong et al. [56] suggests that different amplitude of seasonal trend is caused by stratified atmospheric delay at different elevations. These two factors may both exist and influence the seasonal amplitude. To facilitate numerical modeling, we set different rainfall intensity at different elevations to calibrate the stratification effect of seasonal deformation. The simulated seasonal movement of S15 is present in **Figure 9(c)**. In this study, the elevation of S15 was lower than the reference point (**Figure 4**) and thus, the seasonal movements showed an opposite variance. Body loads dominated cumulative downhill deformation among the three conditions (**Figure 9(g)**), and the time series deformation showed a continuous trend (**Figure 9(d)**). Contrary to the accelerated deformation of S12, P1 and P2 in S15 showed a decelerated trend from July 2018, suggesting that the slope tends to be stable. This is the result of constructing retaining walls (**Figure 8(b)**). In numerical modeling, we configured the retaining walls from July 2018. The retaining walls imposed force opposite to body loads and prevented landslide movements [67]. Theoretically, retaining walls cannot cause deformation. To facilitate modeling, we assumed that it caused deformation opposite to downhill deformation. In this sense, retaining wall-induced force became a transient boundary condition. For S12, the new building construction aggravated downhill movements in time series. For S15, the deformation induced by new retaining walls counteracted downhill deformation induced by body loads, and thus, the total time series deformation tended to be stable after July 2018.

5. Conclusions

On slopes in urban areas, landslides usually have complex loading characteristics, including those influenced by natural and human factors. In this chapter, we reveal the landslide motion in Daguan County through the coupling of InSAR and numerical modeling methods. The coupled approach makes it possible for us to derive full-scale motions. We summarize the main findings as follows:

1. The scatterers detected from SAR images are mainly located in the buit-up areas of Daguan County. Because of the geometric distortion and decorrelation, the scatterers were relatively sparse at lower and higher elevations. 24 of 31 landslides were identified successfully based on the InSAR measurement results. Besides, there was good agreement between InSAR and GNSS and inclinometer data, and the few large errors may have been selected for comparison due to inconsistencies between them. In general, the InSAR measurement results could reflect the landslide movement of this area accurately.

2. By coupling InSAR measurement, we used GA to calculate the optimal material properties. The objective function was minimized to $1.3mm/yr$. The numerical modeling described the full-scale landslide motion from surface to bottom, which could help us to understand the physical process of landslide. The simulated results implicated that the the head of S12 suffered a maximum cumulative deformation because of the external loading caused by buildings of Daguan No.1 Middle School. The maximum deformation of S15 occurred at the toe. This may be due to the reduction of the effective modulus of elasticity of the stratum after soil washing.

3. In the numerical modeling, three boundary conditions are set as external loading. It becomes feasible to evaluate the independent effects of each boundary condition. The results showed that the main load dominates the

cumulative deformation of the landslide by squeezing the water and air in the void. Precipitation induced seasonal motion in a direction perpendicular to the slope surface, and this seasonal feature only appeared in the ascending SAR images. The human activities led to the permanent changes in the time series trend, both positive and negative effects. In particular, the reconstruction projection of Genyun Building accelerated the downhill movement in S12 after May 2018. In S15, the construction of the retaining walls applying a force opposite to the body load, which reduced the downhill movement after July 2018.

Acknowledgements

The authors would like to thank European Space Agency (ESA) for providing free and open Sentinel-1 data. This work was supported by the National Natural Science Foundation of China (41971278, 41601356, 42077238, 41941019), the Research Grants Council (RGC) of Hong Kong (CUHK14504219) and the Second Tibetan Plateau Scientific Expedition and Research Program (2019QZKK0903).

Author details

Peifeng Ma[1*], Yifei Cui[2], Weixi Wang[3], Hui Lin[4], Yuanzhi Zhang[5] and Yi Zheng[1]

1 Institute of Space and Earth Information Science, The Chinese University of Hong Kong, Hong Kong, China

2 State Key Laboratory of Hydroscience and Engineering, Tsinghua University, Beijing, China

3 Research Institute for Smart Cities, School of Architecture and Urban Planning, Shenzhen University, Shenzhen, China

4 School of Geology and Environment, Jiangxi Normal University, Nanchang, China

5 University of Chinese Academy of Sciences, Beijing, China

*Address all correspondence to: peifengma@cuhk.edu.hk

IntechOpen

References

[1] Haque U, Blum P, Da Silva PF, Andersen P, Pilz J, Chalov SR, et al. Fatal landslides in Europe. Landslides. 2016;**13**(6):1545-1554

[2] Intarawichian N, Dasananda S. Frequency ratio model based landslide susceptibility mapping in lower Mae Chaem watershed, Northern Thailand. Environmental Earth Sciences. 2011; **64**(8):2271-2285

[3] Liu D, Cui Y, Guo J, Yu Z, Chan D, Lei M. Investigating the effects of clay/sand content on depositional mechanisms of submarine debris flows through physical and numerical modeling. Landslides. 2020;**17**(8): 1863-1880

[4] Pazzi V, Morelli S, Fanti R. A review of the advantages and limitations of geophysical investigations in landslide studies. International Journal of Geophysics. 2019;**2019**

[5] Strozzi T, Klimeš J, Frey H, Caduff R, Huggel C, Wegmüller U, et al. Satellite SAR interferometry for the improved assessment of the state of activity of landslides: A case study from the cordilleras of Peru. Remote Sensing of Environment. 2018;**217**:111-125

[6] Wang Q, Li W, Yan S, Wu Y, Pei Y. GIS based frequency ratio and index of entropy models to landslide susceptibility mapping (Daguan, China). Environmental Earth Sciences. 2016;**75**(9):780

[7] Cui Y, Cheng D, Choi CE, Jin W, Lei Y, Kargel JS. The cost of rapid and haphazard urbanization: Lessons learned from the Freetown landslide disaster. Landslides. 2019;**16**(6): 1167-1176

[8] Fan X, Scaringi G, Korup O, West AJ, van Westen CJ, Tanyas H, et al. Earthquake-induced chains of geologic hazards: Patterns, mechanisms, and impacts. Reviews of Geophysics. 2019; **57**(2):421-503

[9] Guo C, Cui Y. Pore structure characteristics of debris flow source material in the Wenchuan earthquake area. Engineering Geology. 2020;**267**: 105499

[10] Guzzetti F, Mondini AC, Cardinali M, Fiorucci F, Santangelo M, Chang KT. Landslide inventory maps: New tools for an old problem. Earth-Science Reviews. 2012;**112**(1–2):42-66

[11] Rosi A, Tofani V, Tanteri L, Stefanelli CT, Agostini A, Catani F, et al. The new landslide inventory of Tuscany (Italy) updated with PS-InSAR: Geomorphological features and landslide distribution. Landslides. 2018;**15**(1):5-19

[12] Cigna F, Bianchini S, Casagli N. How to assess landslide activity and intensity with persistent Scatterer interferometry (PSI): The PSI-based matrix approach. Landslides. 2013; **10**(3):267-283

[13] Crosetto M, Monserrat O, Cuevas-González M, Devanthéry N, Crippa B. Persistent scatterer interferometry: A review. ISPRS Journal of Photogrammetry and Remote Sensing. 2016;**115**:78-89

[14] Massonnet D, Rossi M, Carmona C, Adragna F, Peltzer G, Feigl K, et al. The displacement field of the landers earthquake mapped by radar interferometry. Nature. 1993; **364**(6433):138-142

[15] Bonì R, Bordoni M, Vivaldi V, Troisi C, Tararbra M, Lanteri L, et al. Assessment of the Sentinel-1 based ground motion data feasibility for large scale landslide monitoring. Landslides. 2020;**17**:2287-2299

[16] Rucci A, Ferretti A, Guarnieri AM, Rocca F. Sentinel 1 SAR interferometry applications: The outlook for sub millimeter measurements. Remote Sensing of Environment. 2012;**120**:156-163

[17] Ferretti A, Fumagalli A, Novali F, Prati C, Rocca F, Rucci A. A new algorithm for processing interferometric data-stacks: SqueeSAR. IEEE Transactions on Geoscience and Remote Sensing. 2011;**49**(9):3460-3470

[18] Ferretti A, Prati C, Rocca F. Permanent scatterers in SAR interferometry. IEEE Transactions on Geoscience and Remote Sensing. 2001;**39**(1):8-20

[19] Ma P, Wang W, Zhang B, Wang J, Shi G, Huang G, et al. Remotely sensing large-and small-scale ground subsidence: A case study of the Guangdong–Hong Kong–Macao Greater Bay Area of China. Remote Sensing of Environment. 2019;**232**:111282

[20] Bianchini S, Raspini F, Solari L, Del Soldato M, Ciampalini A, Rosi A, et al. From picture to movie: Twenty years of ground deformation recording over Tuscany region (Italy) with satellite InSAR. Frontiers in Earth Science. 2018;**6**:177

[21] Hilley GE, Bürgmann R, Ferretti A, Novali F, Rocca F. Dynamics of slow-moving landslides from permanent scatterer analysis. Science. 2004;**304**(5679):1952-1955

[22] Intrieri E, Raspini F, Fumagalli A, Lu P, Del Conte S, Farina P, et al. The Maoxian landslide as seen from space: Detecting precursors of failure with Sentinel-1 data. Landslides. 2018;**15**(1):123-133

[23] Righini G, Pancioli V, Casagli N. Updating landslide inventory maps using persistent Scatterer interferometry (PSI). International Journal of Remote Sensing. 2012;**33**(7):2068-2096

[24] Singh LP, Van Westen C, Ray PC, Pasquali P. Accuracy assessment of InSAR derived input maps for landslide susceptibility analysis: A case study from the Swiss Alps. Landslides. 2005;**2**(3):221-228

[25] Solari L, Bianchini S, Franceschini R, Barra A, Monserrat O, Thuegaz P, et al. Satellite interferometric data for landslide intensity evaluation in mountainous regions. International Journal of Applied Earth Observation and Geoinformation. 2020;**87**:102028

[26] Hu X, Bürgmann R, Schulz WH, Fielding EJ. Four-dimensional surface motions of the Slumgullion landslide and quantification of hydrometeorological forcing. Nature Communications. 2020;**11**(1):1-9

[27] Kavanagh JL, Engwell SL, Martin SA. A review of laboratory and numerical modelling in volcanology. Solid Earth. 2018;**9**(2):531-571

[28] Smith IM, Griffiths DV, Margetts L. Programming the Finite Element Method. Chennai, India: John Wiley & Sons; 2013

[29] Zienkiewicz OC, Taylor RL, Taylor RL, Taylor RL. The Finite Element Method: Solid Mechanics. Vol. 2. Barcelona, Spain: Butterworth-Heinemann; 2000

[30] Zhou W, Li S, Zhou Z, Chang X. Insar observation and numerical modeling of the earth-dam displacement of shuibuya dam (China). Remote Sensing. 2016;**8**(10):877

[31] Ma P, Cui Y, Wang W, Lin H, Zhang Y. Coupling InSAR and numerical modeling for characterizing landslide movements under complex loads in urbanized hillslopes. Landslides. 2021;**18**(5):1611-1623

[32] Chen X, Zhou Q, Ran H, Dong R. Earthquake-triggered landslides in

Southwest China. Natural Hazards and Earth System Sciences. 2012;**12**(2):351-363

[33] Liu X, Wang S, Zhang X. Influence of geologic factors on landslides in Zhaotong, Yunnan province, China. Environmental Geology and Water Sciences. 1992;**19**(1):17-20

[34] Siles G, Trudel M, Peters DL, Leconte R. Hydrological monitoring of high-latitude shallow water bodies from high-resolution space-borne D-InSAR. Remote Sensing of Environment. 2020;**236**:111444

[35] Yan G, Yin Y, Huang B, Zhang Z, Zhu S. Formation mechanism and characteristics of the Jinjiling landslide in Wushan in the three gorges reservoir region, China. Landslides. 2019;**16**(11): 2087-2101

[36] Ma P, Lin H. Robust detection of single and double persistent scatterers in urban built environments. IEEE Transactions on Geoscience and Remote Sensing. 2015;**54**(4):2124-2139

[37] Ma P, Liu Y, Wang W, Lin H. Optimization of PSInSAR networks with application to TomoSAR for full detection of single and double persistent scatterers. Remote Sensing Letters. 2019;**10**(8):717-725

[38] Shi G, Lin H, Bürgmann R, Ma P, Wang J, Liu Y. Early soil consolidation from magnetic extensometers and full resolution SAR interferometry over highly decorrelated reclaimed lands. Remote Sensing of Environment. 2019; **231**:111231

[39] Ferretti A, Prati C, Rocca F. Nonlinear subsidence rate estimation using permanent scatterers in differential SAR interferometry. IEEE Transactions on Geoscience and Remote Sensing. 2000;**38**(5):2202-2212

[40] Stoica P, Moses RL. Introduction to spectral analysis. Pearson. Education. 1997

[41] Zhu XX, Bamler R. Tomographic SAR inversion by $L\_\{1\}$-norm regularization—The compressive sensing approach. IEEE Transactions on Geoscience and Remote Sensing. 2010; **48**(10):3839-3846

[42] Colesanti C, Ferretti A, Novali F, Prati C, Rocca F. SAR monitoring of progressive and seasonal ground deformation using the permanent scatterers technique. IEEE Transactions on Geoscience and Remote Sensing. 2003;**41**(7):1685-1701

[43] Huber PJ. Robust estimation of a location parameter. In: Breakthroughs in Statistics. New York, NY: Springer; 1992. pp. 492-518

[44] Hansen PC, O'Leary DP. The use of the L-curve in the regularization of discrete ill-posed problems. SIAM Journal on Scientific Computing. 1993; **14**(6):1487-1503

[45] Zhang B, Wang R, Deng Y, Ma P, Lin H, Wang J. Mapping the Yellow River Delta land subsidence with multitemporal SAR interferometry by exploiting both persistent and distributed scatterers. ISPRS Journal of Photogrammetry and Remote Sensing. 2019;**148**:157-173

[46] Guarnieri AM, Tebaldini S. On the exploitation of target statistics for SAR interferometry applications. IEEE Transactions on Geoscience and Remote Sensing. 2008;**46**(11):3436-3443

[47] Rucci A, Vasco D, Novali F. Monitoring the geologic storage of carbon dioxide using multicomponent SAR interferometry. Geophysical Journal International. 2013;**193**(1):197-208

[48] Hanssen RF. Radar Interferometry: Data Interpretation and Error Analysis. Vol. 2. Dordrecht: Springer Science & Business Media; 2001

[49] Krahn J. Stress and Deformation Modeling with SIGMA/W. Alberta,

Canada: GEO–SLOPE International, Ltd; 2004

[50] Segerlind LJ. Applied Finite Element Analysis. Vol. 316. New York: Wiley; 1976

[51] Zhang J, Tang WH, Zhang L. Efficient probabilistic back-analysis of slope stability model parameters. Journal of Geotechnical and Geoenvironmental Engineering. 2010; **136**(1):99-109

[52] Houck CR, Joines J, Kay MG. A genetic algorithm for function optimization: A Matlab implementation. Ncsu-ie tr. 1995;**95**(09):1-10

[53] Kulhawy FH, Mayne PW. Manual on Estimating Soil Properties for Foundation Design. Cornell Univ., Ithaca ...: Electric Power Research Inst., Palo Alto, CA (USA); 1990

[54] Rawls WJ, Brakensiek D. Estimating soil water retention from soil properties. Journal of the Irrigation and Drainage Division. 1982;**108**(2):166-171

[55] Ritter A, Hupet F, Muñoz-Carpena R, Lambot S, Vanclooster M. Using inverse methods for estimating soil hydraulic properties from field data as an alternative to direct methods. Agricultural Water Management. 2003; **59**(2):77-96

[56] Dong J, Zhang L, Liao M, Gong J. Improved correction of seasonal tropospheric delay in InSAR observations for landslide deformation monitoring. Remote Sensing of Environment. 2019;**233**:111370

[57] Hu X, Wang T, Pierson TC, Lu Z, Kim J, Cecere TH. Detecting seasonal landslide movement within the Cascade landslide complex (Washington) using time-series SAR imagery. Remote Sensing of Environment. 2016;**187**:49-61

[58] Hashash Y, Jung S, Ghaboussi J. Numerical implementation of a neural network based material model in finite element analysis. International Journal for Numerical Methods in Engineering. 2004;**59**(7):989-1005

[59] Bonì R, Bordoni M, Colombo A, Lanteri L, Meisina C. Landslide state of activity maps by combining multi-temporal A-DInSAR (LAMBDA). Remote Sensing of Environment. 2018; **217**:172-190

[60] Coe JA, Ellis WL, Godt JW, Savage WZ, Savage JE, Michael J, et al. Seasonal movement of the Slumgullion landslide determined from global positioning system surveys and field instrumentation, July 1998–march 2002. Engineering Geology. 2003;**68**(1–2): 67-101

[61] Cui Y, Chan D, Nouri A. Coupling of solid deformation and pore pressure for undrained deformation—A discrete element method approach. International Journal for Numerical and Analytical Methods in Geomechanics. 2017;**41**(18): 1943-1961

[62] Handwerger AL, Roering JJ, Schmidt DA. Controls on the seasonal deformation of slow-moving landslides. Earth and Planetary Science Letters. 2013;**377**:239-247

[63] Zhao C, Lu Z, Zhang Q, de La Fuente J. Large-area landslide detection and monitoring with ALOS/PALSAR imagery data over northern California and southern Oregon, USA. Remote Sensing of Environment. 2012;**124**: 348-359

[64] Attanayake PM, Waterman MK. Identifying environmental impacts of underground construction. Hydrogeology Journal. 2006;**14**(7): 1160-1170

[65] Thomson S, Tiedemann C. A review of factors affecting landslides in urban areas. Bulletin of the Association of

Engineering Geologists. 1982;**19**(1): 55-65

[66] Wong H, Ho K. The 23 July 1994 landslide at Kwun lung Lau, Hong Kong. Canadian Geotechnical Journal. 1997; **34**(6):825-840

[67] Yingren Z, Shangyi Z. Calculation of inner force of support structure for landslide/slope by using strength reduction FEM [J]. Chinese Journal of Rock Mechanics and Engineering. 2004;**20**

Section 6

Landslide Inventory

Chapter 11

Landslide Inventory, Susceptibility, Hazard and Risk Mapping

Azemeraw Wubalem

Abstract

Landslide is that the downslope movement of debris, rocks, or earth material under the influence of the force of gravity. Although the causes and mechanisms of landslides are complicated, human action, earthquakes, and severe rainfall can trigger them. It can happen when the driving force surpasses the resisting force due to natural soil or rock slope destabilization. Landslide is one of the foremost destructive and dangerous natural hazards that cause numerous fatalities and economic losses worldwide. Therefore, landslide investigation, susceptibility, hazard, and risk mapping are vital tasks to disaster loss reduction and performance as a suggestion for sustainable land use planning. The determination of the cause variables, identification of existing landslides, and production of a landslide susceptibility, hazard, and risk map are all necessary steps in the mitigation of landslide incidence on the globe. Landslide susceptibility, hazard, and risk maps are the outcome of a statistical relationship between environmental conditions and previously occurring landslides. It provides critical scientific support for the government's reaction to land use practices and the management of landslide threats. The type, concept of landslides, factor, inventories, susceptibility, hazard, and risk, as well as mapping and validation methodologies, have all been examined in this chapter. The distinction between landslide susceptibility and hazard has surely been debated.

Keywords: susceptibility, hazard, risk, inventory

1. Introduction

Landslide inventory, susceptibility, hazard, and risk mapping may be a complex job thanks to a good spectrum of conditioning and triggering factors, lack of record data, and non-uniqueness of mapping methods. As a result, a geologist's participation in landslide inventory, susceptibility, hazard, and risk mapping is critical. In landslide susceptibility, hazard, and risk mapping, mapping and analysis of previous and active landslide incidence are demanding tasks that can be used for landslide prevention and mitigation. Landslide disaster prevention and mitigation will not be effective unless the landslide-prone area is correctly mapped [1]. Landslides can bury animals and persons; demolish houses, farms, and infrastructures in a short amount of time [2] and Wubalem [3]. Hong et al. [2], Wubalem [4] are stated that within a short period, landslides can bury animals and humans, destroy houses, farms, and infrastructures. Landslide is one of the foremost destructive and dangerous natural hazards that cause numerous fatalities and economic losses worldwide [2, 5–7]. Therefore, landslide inventory, susceptibility, hazard, and risk mapping and assessment are vital to disaster loss reduction and function as a suggestion for sustainable land use planning.

The extenuation actions of landslide incidence within the planet are required determination of the causal factors, identification of prevailing landslides, and generation of landslide susceptibility, hazard, and risk map [8]. Landslide inventory mapping is extremely important to work out landslide type, failure mechanism, spatial distribution, and size in a given region. Landslide inventory is also important for landslide susceptibility, hazard, and risk mapping. Chen and Wang [9] explained that susceptibility, hazard, and risk maps of landslides are the results of the statistical relationship in between landslide governing factors and preexisting landslides. Susceptibility, hazard, and risk map of landslides are imperative for scientific support of the government's response to land use practice and landslide hazards management [9, 10]. The landslide susceptibility or hazard mapping is not only to determine the factors that are most influential to the landslides that occurred within the region but also to appraisal the comparative influence of every landslide governing factors [9]. As stated by Chen and Wang [9], landslide susceptibility or hazard mapping is also significant to inaugurate an association between the factors and landslides to foresee the landslide hazard in the future. As a result, extensive and accurate landslide inventory mapping, as well as the creation of landslide susceptibility, hazard, and risk maps, is critical. Although the reason for landslide incidence and its mechanisms are so complex, human interventions, earthquakes, and heavy rainfall can trigger it. As Kifle [11]; Wubalem and Meten [4] stated that landslide incidence can also occur when the resistance force exceeds by driving force thanks to the destabilization of natural soil or rock slopes. This chapter is provided a summary of the sort of landslide type, factor, landslide inventory, landslide susceptibility, hazard, risk mapping, and validation approaches.

2. Definition and concepts

Landslide is that the movement of the mass of rock, debris, and earth downslope [12–15]. Landslides are also defined as an outsized range of geotechnical phenomena under the influence of gravity. On another hand, a landslide is that the type of mass wasting activity that denotes any outward or downslope movement of soil and rock under the direct influence of gravity when the drive exceeds the resistance force of a slope [13, 14, 16]. These masses may range in size from card to entire mountainsides. Their movements may vary in velocities. Landslide as a geological hazard is caused by earthquake or eruption, rainfall, and act. This is often initiated when an area of a hill slope or sloping section of the seabed is rendered weak to support its weight. It is one of the foremost destructive natural hazards triggered by natural and man-made factors like an earthquake, rainfall [17], and act like an improper/poor quarry, and road construction/inadequate maintenance in mountainous terrain [18].

In geohazard mapping, susceptibility/vulnerability, hazard, and risk mapping are the foremost important activities to understand, mapping, and evaluating the spatiotemporal condition and level of risk because of geo-hazards. These terms have different meanings but some researchers use the terms interchangeably. Susceptibility refers to the probability of occurrence of an event within a selected type during a given location whereas hazard refers to the probability of occurrence of an event within a selected type and magnitude during a given location within a reference period. This means, susceptibility is usually used to predict the spatial occurrence of events, but the hazard is usually used to predict the spatiotemporal occurrence of events during a given terrain. The term risk refers to the expected losses or damage by events during a given region, which are the products of susceptibility, hazard, and elements in peril. Vulnerability means the degree of loss to a given element of the set of elements in peril resulting from the occurrence of natural phenomena of a given magnitude. It is expressed on a scale from 0 (no damage) to 1 (total damage). Elements at risk is potentially vulnerable of properties, population, and economic activities including public services in peril during a given area.

2.1 Type of landslide or failure mechanism

Landslides are usually classified based on the materials involved (rocks, debris, and soils) and on their mechanism and failure (**Table 1**). Other factors include groundwater content and the rate and dimension of the movement. Classifying and studying this phenomenon is important to manage damages because of the landslide. Classification of the landslide is the primary step to investigate landslides. According to Varnes [13, 19], landslides are classified based on the types of material, mode of movement, landslide activity, the rate of movement, depth, the magnitude of slide and moisture content.

2.1.1 Rotational landslides

Rotational landslides are more common in cohesive, homogeneous soils. The failure, which can be superficial or deep-rooted, occurs along curved surfaces concave upwards, having a shape of a spoon. Successive landslides occur mainly in stiff fissured clays with gradients similar to their angle of equilibrium and in soft very sensitive clays, where the initial landslide causes an accumulation of remolded clay, which as it flows, leaves the material higher up without support, so promoting successive failures. These failures are shallow but can have considerable lateral continuity [20]. Weak rock masses or those with a high degree of fracturing or weathering, where the structural discontinuities do not form preferred surfaces for failure may also suffer this type of successive landslides.

2.1.2 Translational slides

In translational slides, failure takes place along pre-existing planar surfaces or discontinuities (bedding planes, contact between different types of materials,

Movement type		Slope material type			Source
		Bedrock	Soil mass in failed slope		[13]
			Principal coarse	Principal fine	[13]
Topples		Rock topple	Debris topple	Earth topple	[13]
Fall		Rock fall	Debris fall	Earth fall	[13]
Lateral spread		Rock spread	Debris spread	Earth spread	[13]
Slide	Translational slide	Rockslide	Debris slide	Earth slide	[13]
	Rotational slide				
Flows		Rock flow (Deep creep)	Debris flow	Earth flow	[13]
Composite or complex Two or more principal types of movement in combination					[13]

Table 1.
Landslide classifications based on material and types of movements [13].

structural surfaces, etc.) and sometimes the failure plane is a fine layer of clay material between more competent strata [20].

The sliding mass can be sometimes rectangular blocks that have been detached from the mass at discontinuities or tension cracks (block landslides). Translational slides generally move faster than rotational ones, because of their simple geometry of failure mechanism.

2.1.3 Flows

As defined by Vallejo and Ferrer [20], flows are mass movements of soil (mud or earth flows), debris (debris flows), or rock blocks (rock fragment flows) often with high water content, where the material behaves as a fluid undergoing continuous deformation but without having well-defined failure surfaces. Water is the main triggering factor because water decreases the strength of materials having low cohesion [20]. Flows mainly affect sensitive clay soils which show considerable loss of strength when mobilized; these movements are not very deep in their extent and develop on slopes <10°.

2.1.3.1 Mud or earth flows

Mud or Earth flows occur in predominantly fine and homogeneous materials and may move at a speed of the many meters per second; the loss of strength is typically caused by water saturation. They are classified consistent with the sort of fabric, its strength, and its water content. Mudflows are generally small-scale and slow but sometimes especially in-saturated conditions, they are extensive and fast, with catastrophic consequences once they reach populated areas. Fine volcanic materials are particularly vulnerable to this sort of process.

2.1.3.2 Debris flows

Debris flows are complex movements, which include rock fragments, blocks, cobbles, and gravel in a fine-grained matrix of sands, silts, and clays. They occur on slopes covered with loose or non-consolidated material, especially where there is no vegetation cover.

2.1.4 Creep

Creep may be a very slow, almost imperceptible superficial movement (a few decimeters deep), which affects soils and weathered materials, causing continuous deformations that becomes progressively noticeable on slopes over time. This causes fences, walls, or posts to lean or offset and trees to be bent. Creep may be a time-dependent deformation and defines the deformational behavior of the fabric instead of the sort of movement.

2.1.5 Solifluction

Solifluction affects the saturated surface layer of slopes. This is often a slow movement produced by the freeze–thaw process because the daily or seasonal temperature variations change the water phase and water content of fine-grained soils in cold regions.

2.1.6 Rock falls

Rock falls are very quick free falls of rocks, which are dislodged from pre-existing discontinuity planes (tectonic, bedding surfaces, and tension cracks). The movement could also be by a vertical fall, by a series of bounces, or by rolling down the slope surface. They are common on steep slopes in mountainous areas, on cliffs, and generally, on rock walls and therefore the blocks are bounded by different sets of discontinuities often forming wedge-shaped blocks. The factors that cause rock falls include erosion and loss of support for previously loosened blocks in steep slopes, water pressures in discontinuities, and tension cracks and seismic shakes. Although the fallen blocks could also be relatively small in terms of volume, rock falls are sudden processes that pose a big risk to communication routes and buildings in mountainous zones and at the foot of steep slopes. Masses of soil can also fall from vertical natural and excavated slopes, thanks to the existence of tension cracks generated by tensional stresses or shrinkage cracks within the ground that has dried.

2.1.7 Toppling

The toppling of strata or blocks of rock may be included in rock falls. Toppling occurs when the strata dip in the opposite direction to the slope and form naturally inclined blocks, which are free to rotate because of failure at the foot of the slope. Toppling tends to occur mainly on rocky slope faces, which intersect steeply dipping strata [20].

2.1.8 Rock avalanches

Rock avalanches are rapidly falling masses of rock and debris that detach themselves from steep slopes, sometimes amid ice or snow. The rock masses disintegrate during their fall and form deposits of very different block sizes and form deposits of very different block sizes, with no rounding from abrasion and chaotic distribution [20]. Rock avalanche deposits are unstructured and have great porosity [20]. Avalanches are generally the results of large-scale landslides or rock falls during which due to the steep gradient and therefore the lack of both structure and cohesion in their materials, travel down over steep slopes at great speed (up to 100 Km/h).

2.1.9 Debris avalanches

Debris avalanches are formed from rock material containing an excellent sort of sizes and should include large blocks and abundant fines [20]. Loose deposits and loose materials resulting from volcanic eruptions are susceptible to this process. The most difference with debris flows, aside from water content (which is not necessary for debris avalanches), is that the rate and speed of movement of the avalanche in areas of a steep gradient.

2.1.10 Lateral displacements

This sort of movement (also called lateral spreading) refers to the movement of rock blocks or coherent, cemented soil masses that rest on soft & deformable slopes. These movements are thanks to the loss of strength of the underlying material, which either flows or deformed under the load of the rigid blocks. Lateral spreading can also cause by liquefaction of the underlying material or by lateral extrusion of sentimental, wet clays under the load of the masses above them [20]. These movements occur on gentle slopes and should be very extensive.

2.2 States and distribution of landslide

Determining the states and distribution of landslides is extremely important to repair the consequences of landslides on infrastructures, lives, farmlands, and environments. The landslide are going to be found within the following different states of condition. Active landslide is currently moving. A suspended landslide has moved within the last twelve months but is not active at the present. A reactivated landslide is a lively landslide that has been inactive. An inactive landslide is a landslide, which did not moved at most for year.

Inactive landslides are often subdivided into these states:

- A dormant landslide is an inactive landslide, which will be reactivated by its original causes or other causes.

- An abandoned landslide is an inactive landslide that is not suffering from its original causes.

- A stabilized landslide is an inactive landslide that has been shielded from its original causes by artificial remedial measures.

- A relict landslide is an inactive landslide that developed under geomorphological or climate considerably different from those at the present.

2.3 Recognition of landslides

Potential and existed landslides can be identified or recognized using different techniques considering various features that existed on the earth's surface. Different features indicate landslide signs like

- Depression at top (water ponding)

- Bulging at toe Tension cracks

- Water seepage (generally at toe)

- Tilted and crooked trees

- Change in vegetation

- Change in topography

- Change in drainage pattern

2.4 Landslide factors

2.4.1 Introduction

In hazard minimization, the evaluation of landslide conditioning and triggering factors is a very important task. Geodynamic processes affecting the earth's surface cause mass movements of different types, sizes, and speeds [20]. Landslide movement is that the most frequent and widespread sort of mass movement generated by the gravitational downslope displacement of soil and rock masses [20]. The force of gravity and therefore the progressive weakening of geological materials, mainly thanks to weathering, alongside the action of other natural and environmental phenomena, make mass movements relatively common on the earth's surface [20]. These processes create potential geological risks, as they will cause economic loss and social damage if they affect human activities, buildings, and infrastructure [20]. How to avoid these adverse effects is the subject of research including mass movements, their characteristics, instability mechanisms, controlling factors, and causes. To carry out this research, it is necessary to understand the characteristics and therefore the geological, geotechnical, and hydrogeological properties of the soil and rock materials involved and their mechanical behavior also because the factors that condition and trigger such movements [20]. Studies during this field should specialize in the investigation of [20]

- Particular processes for the design of stabilizing measures to either mitigate or reduce damage.

- Analysis of the factors, which control and trigger processes at particular locations, to stop possible movements.

- Mapping either unstable or potentially unstable zones, in order that the hazardous areas are often delimited and preventive measures are often applied.

As usual, landslides might transpire when shear stress exceeds the shear strength of slope material. The factors that cause landslide have been classified as factors that contribute to an increase of the shear stress and factors that contribute to the decrease of shear strength; however, water is another factor contributing to both increasing and decreasing shear stress and shear strength of slope material respectively. Factors these increase shear stresses are included removal of lateral support; surcharge/ overloading, transitory earth stress, regional tilting, removal of underline support, and increase in lateral pressure. The factors that contribute to the decrease of shear strength of slope material include factors like initial state or inherent characteristics of materials and the changing or variable factors that tend to lower the shear strength of a material. On other hand, factors that control landslides are classified into two such as intrinsic/inherent/static and external/ dynamic landslide factors [21–23].

2.4.2 Intrinsic controlling factors

According to Anbalangan [21], and Raghuvashi et al. [24], intrinsic parameters are the inherent controlling factors that outline the favorable or unfavorable condition within the slope. These include slope material, slope geometry, structural discontinuity, land use/cover, and groundwater. These factors have an excellent influence to decrease the strength of the slope material. Hence, mapping and perception of their impression are crucial for slope stability analysis.

2.4.2.1 Lithology

The kind of fabric during a slope is closely associated with the sort of instability. Different lithology are going to be showed different degrees of susceptibility to potential slippage or failure. The stress–strain behavior of materials is governed by their strength properties, which also depend upon the presence of water. Sorts of failure and therefore the location of failure surfaces depend upon factors like alternating materials of various lithology, the extent of weathering, and therefore the presence of layers of sentimental material or hard strata. Soils, which are considered homogeneous materials, compared to rock masses, instability could also be generated by differences within the degree of compaction, cementation, and grain size, which can make sure areas more vulnerable to weakness and water flow. In rock masses, characterization and analysis of slope behavior are further complicated by the presence of layers of strata with differing strengths and properties [20].

2.4.2.2 Discontinuities

Geological structures or discontinuities play a definitive role in conditioning the slope stability in rock masses. A mixture of structural elements and geometric slope parameters, like height, gradient, and orientation, defines problems, which will occur. The spatial distribution of discontinuities is that the structure of the rock mass [20]. The presence of those surfaces of weakness (bedding surfaces, joints, and faults) dipping towards the slope face implies the existence of potential failure planes on which sliding can readily occur.

The orientation and spatial distribution of discontinuities will condition the sort and mechanism of the instability. A specific system of fracturing will condition both the direction of movement and therefore the size of blocks susceptible to slide or the presence of a fault dipping towards a slope face will limit the unstable area. Structural changes and singularities within the rock mass, like Tectonized or shear areas, or abrupt changes within the dip of the strata, indicate heterogeneities from which failure might originate. Slope stability could also be suffering from changes to the initial conditions during excavation; for instance, the existence of tectonic in place stress related to compressive or extensional structures like folds and faults.

2.4.2.3 Hydrogeological conditions

Most failures are caused by the effects of water in the ground, including pore pressures and erosion of the slope materials. Water is considered the worst enemy of slope stability, together with human actions where excavations are carried out without adequate geotechnical care. The presence of water in a slope reduces stability by decreasing ground strength and increasing forces, which favor instability. The main effects of water are a reduction in the shear strength of failure surfaces as effective normal stress, σ'_n, decreases. $\tau = c + \left(\sigma'_n - u\right)\tan\varphi$. Increase in the downslope shear forces as water pressure is exerted in tension cracks. Increase in weight of the material due to saturation: $\gamma = \gamma_d + Sn\gamma_w$ where γ_d = dry apparent unit weight; S = degree of saturation; n = porosity and γ_w = unit weight of water. Softening of soils associated with an increase in their water content. Internal erosion or piping caused by surface or underground flow. The shape of the water table on a slope depends on such factors as the permeability of materials and the geometry or shape of the slope. In rock masses, the configuration of the water table is greatly influenced by the geological structure and the alternation of permeable and impermeable materials, which in turn affect the distribution of pore water pressures on any potential slip surfaces. The influence of water on the properties of materials depends on their hydrogeological behavior. The greatest effect is produced by the pressure exerted defined by the piezo metric head [20].

The following aspects should be known to understand the effects of water in a slope [20]: 1) Hydrogeological behavior of the materials 2) Presence of water table and piezo metric heads 3) Water flow in the slope 4) Relevant hydrogeological parameters:

permeability coefficient or hydraulic conductivity, hydraulic gradient, transmissivity, and storage coefficient. One way of obtaining an approximate assessment of the entire force exerted by water on discontinuity surfaces or tension cracks is to assume the triangular distribution of hydrostatic pressure on these surfaces.

2.4.2.4 Properties of soil and rock masses

The possible failure of a slope along a surface depends on the strength, which depends on cohesion and therefore the interior angle of friction. The influence of geological history (e.g. consolidation, erosion, diagenetic processes, in situ stresses, and weathering) on the mechanical (shear strength) properties of soils must be determined considering the geological characteristics. In rock masses, mechanical behavior is decided by the strength properties of the discontinuities and therefore the intact rock counting on its degree of fracturing and the nature of the materials and discontinuities within it. The behavior of a tough rock mass generally depends on the characteristics of its discontinuities, although the lithology and its geological evolution can also play a crucial role. The shear strength of surfaces of weakness depends on their nature and origin, persistence, spacing, roughness, type and thickness of infill, and thus the presence of water.

Slope stability is highly control by in situ stresses [20]. The strain relief from decompression when the slope is excavated may transform its material properties [20]. In rock slopes, the weakest areas are often degraded and begin to behave like soft rock or granular soil. This effect is common in mudstone or mud-shale slopes subjected to high in place stresses; the rock formation is weakened into a granular material with cement-sized fragments several meters thick inside the slope, resulting in disintegration and collapse of the slope.

2.4.2.5 Geomorphological factors (slope, aspect, and curvature)

Slope morphometry refers to the steepness of the slope, which controls not only the strain distribution inside the slope mass but also affects weathering layer depth and surface runoff [25]. As reported by Lai [25], the degree and height of the slope influence the quantity of runoff and thus the extent of erosion. The steeper the slope, the upper velocity of water flowing down a slope and have higher erosive power. Thus, the slope material that supports the slope are getting to be removed and heighten the slope instability problem.

Aspect is that the orientation of the slope. Different slope direction has different weather, land cover, and radiation intensity that affects the exposure of the slope to radiation, wind impact, and rainfall [26, 27].

Curvature is that the measure of the roughness of a given terrain. The curvature may ask the concaveness, concaveness, and flatness of a slope. According to Pradhan [28, 29]; Alkhasawneh et al. [30], as cited in Meten et al. [26] the negative value refers to the valley, the positive value refers to Capitol Hill slope, and zero/approaches zero value refers to flat acreage. The curvature condition controls the hydraulic condition and thus the consequences of gravity for slope stability.

2.4.3 External triggering factors

External triggering factors are dynamic factors, which may trigger slope movement by increasing driving force. These triggering factors include rainfall, seismic and act. Static and dynamic loads exerted on slopes modify the force distribution and may produce instability. Static loads include the load of structures or buildings on a slope or loads derived from fills, waste dumps, or heavy vehicles, and when these loads are exerted on the slope head, they create a further weight, which will contribute to the destabilizing forces. Dynamic loads are mainly thanks to natural or induced seismicity and vibrations caused by nearby blasting. These mainly affect jointed rock masses by opening up pre-existing discontinuities, reducing their shear strength, and displacing rock blocks, which can then fall. Dynamic forces produced by an action earthquake can be given as a function of the maximum horizontal acceleration. Precipitation and climate regime influence slope stability by modifying groundwater content. The strength of the soil mass becomes loss due to changes in soil structure by alternating periods of rainfall and drought.

Man Made Factors: Abebe et al. [31]; Kifle [11, 32] is explained that the demand for new land for infrastructure, settlement, and agriculture are primary means in which humans can contribute to slope instability condition through the excavation of slope toe or slope faces, loading of slope crest, drawdown (or reservoirs), irrigation, mining, artificial vibration, deforestation, and water leakage from utilities.

2.5 Landslide inventory mapping

Landslide inventory is that the simplest sort of landslide map [33]. The landslide inventory map portrays the spatial distribution, frequency, activity, size, time, type, displace material, the intensity of injury, and density of landslide. It is often used because the base for future landslide susceptibility, hazard, and risk prediction by evaluating the connection between the prevailing landslide event and landslide driving factors [34]. Besides, landslide inventory is often used to evaluate the accuracy and performance of the landslide susceptibility, hazard, and risk maps. Landslide inventory map shows the past and current landslide incidences, which may be prepared using various techniques like the aerial photograph, Google Earth

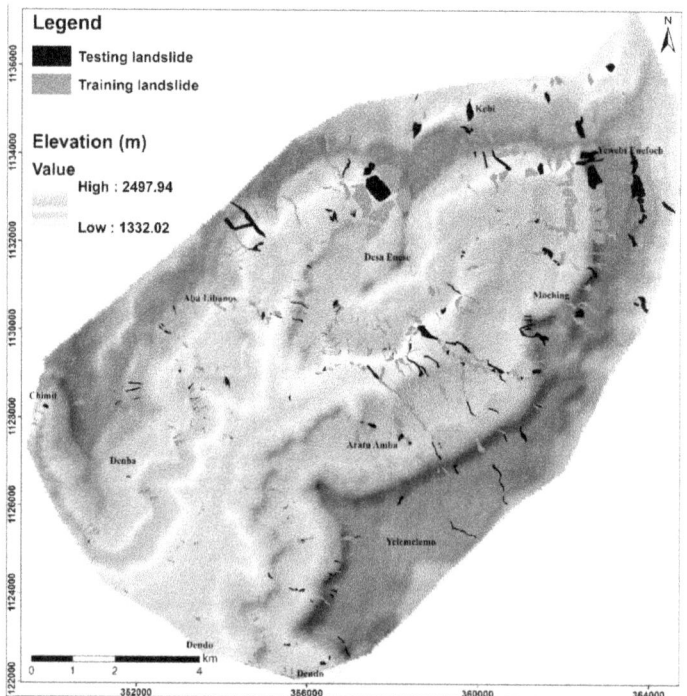

Figure 1.
Landslide inventory map of the study area [36].

imagery, field investigation, and evaluation of archive data including GIS tools. Depend upon the aim, the size of the base map or aerial photograph, the extent of the study area, and therefore the availability of resources, a landslide inventory map are often prepared using different techniques as expressed above [35]. For instance, a small-scale landslide inventory map (1,25,000) landslide inventory maps are often prepared for a selected area using aerial photographs at the size of >1:20,000, Google Earth Imagery analysis, and extensive fieldwork [3, 4, 36, 37]. The Google Earth Imagery may be a free tool that helps not only to spot statistic landslide boundary but also wont to determine the area coverage, perimeter, and distance of slope material movement compared to other techniques, however, it needs field for verification purpose. As a result, currently, from the active and old landslide scarps, researchers intended to spot historical landslides using statistic Google Earth Imagery analysis instead of an aerial photograph. Depend upon the dimensions of

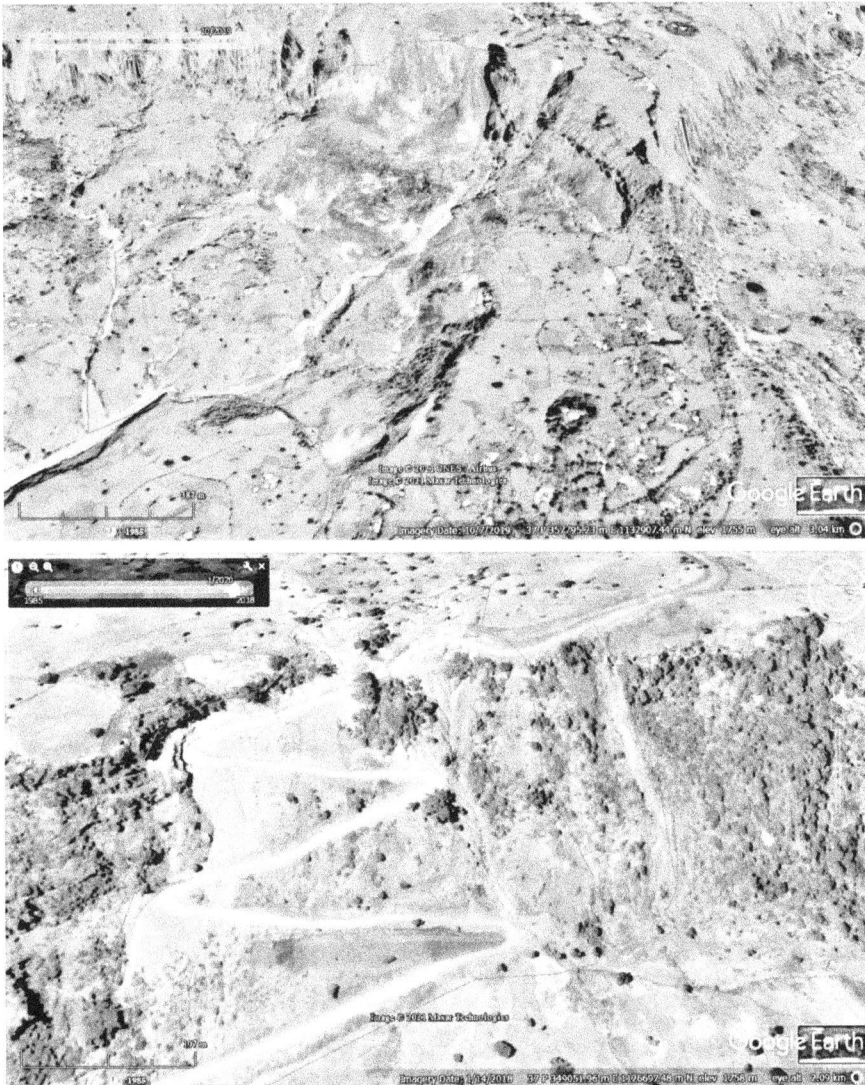

Figure 2.
Landslide in Chemoga catchment, northwestern Ethiopia.

the landslide and therefore the mapping scale, active and old landslide boundaries are often digitized into polygons employing a GIS tool with the assistance of Google Earth Imagery, and eventually, a landslide inventory map are often produced. The landslide inventory is going to be classified as training data sets and testing landslide data sets (**Figure 1**). Most of the researchers classified landslides into 70% for training data sets and 30% for testing landslide data sets [26, 38–40]. As shown in **Figures 2–4**, Google Earth Imagery analysis is so effective for landslide inventory mapping. Landslide investigation is an important task in landslide disaster reduction strategies. It can be conducted to determine and predict old, active, and future landslide incidence by examining land features. For example, field survey is used to evaluate slope gradient, geomorphology, geology, drainage, nature of soil, land use land cover, surface and subsurface water, geodynamic process, old and active landslide conditions. Generally, the methods or techniques that used to investigate landslides are summarized in **Table 2**.

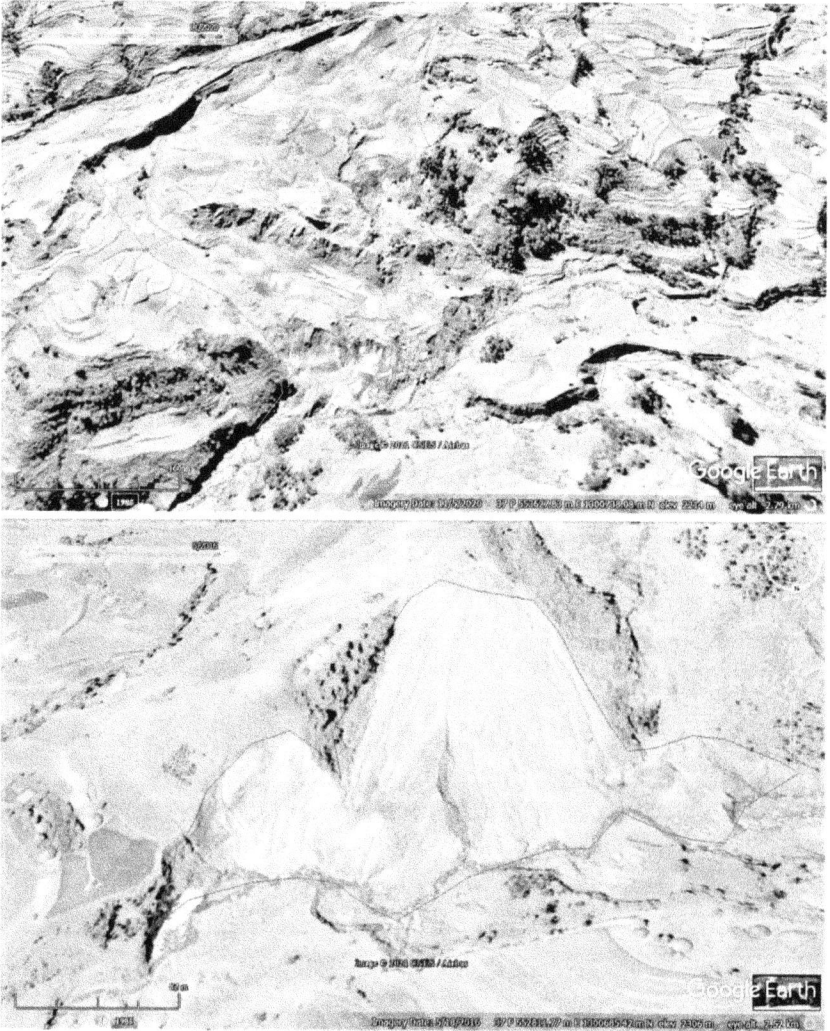

Figure 3.
Landslide in Woldia area, northwestern Ethiopia.

Figure 4.
Landslide in Dessie town, Ethiopia.

Scop	Phase of study	Methods or thechniques	Objectives
Regional landslide study	Preliminary	Review of existing information and existing maps. Google Earth Imagery analysis, Interpretation of aerial photos and remote sensing.	Identify processes and type of movements. Identify conditioning factors. General evaluation of stability of the area. Indentify location and boundary of landslide.
	General study	Field observations. Processes mapping. Factors mapping.	
Conducted to investigate landslides or slope failure for specific area	Study of process and causal factors	Field surveys. Preliminary underground investigation: geophysical methods.	Describe and classify processes and materials. Susceptibility analysis based on the existing processes and concurrence of conditioning factors. Record landslide type, location, magnitude, frequency, dimention, damage, and element at risk.
	Detail investigation	Boreholes, geophysical methods, in situ tests, sampling, Laboratory tests.	Describe and classify movements. Collect morphological, geological, hydrogeological and geomechanical data.
	Monitoring	Inclinometers, extensometers, tiltometers, piezometers.	Collect data on speed, direction, stability analysis using Limit equilibrium methods and Stress–strain numerical models. Determine situation of failure planes, water pressures.
	Stability analysis	Limit equilibrium methods. Stress–strain numerical models.	Define failure models and failure mechanisms. Evaluate stability. Design corrective measures.

Table 2.
Summary of landslide investigation techniques [20].

2.6 Landslide susceptibility, hazard and risk mapping

Landslide susceptibility may be a quantitative or qualitative evaluation of landslide occurrence of a specific type in a given location that is wont to predict spatial distribution, classification, and area of existed or potentially prone area [12, 37]. However, a landslide hazard map is employed to predict future spatial and temporal landslide occurrence with a specific type and magnitude. Although both landslide susceptibility and hazard map are different concepts, many researchers are used the terms as interchangeable. The researchers consider their susceptibility map as a hazard map during which magnitude and frequency did not consider in their model generation. The landslide risk map is employed to predict the expected spatial and temporal losses or damage by landslide incidences during a given region, which are the products of susceptibility or vulnerability, hazard, and elements in danger. Although landslide susceptibility, hazard, and risk maps are the results of the connection between landslide events and sets of landslide factors supported expert judgment or statistical analysis, hazard and risk maps become differ by some input parameters. For instance, a landslide hazard map will have additional landslide frequency, and magnitude input parameters whereas for a risk map, both susceptibility and hazard map become input parameters besides, the element in danger. As stated by Wubalem [3], landslide susceptibility and hazard map results from the sum of all weighted landslide factors employing a raster calculator or weighted overlay method in ArcGIS. Compare to landslide susceptibility mapping, landslide hazard mapping required excellent landslide inventories that contain magnitude, date of occurrence, and frequency. The shortage of frequency, date of occurrence, and magnitude of landslide, landslide hazard mapping become a difficult task. Thus, landslide research trends are shifted to landslide susceptibility mapping for the last twenty century. Now a day, thanks to technological advancement, landslide hazard mapping becomes a simple task for that area frequently suffering from landslide incidence. Lithological, geomorphological, geological structure, hydrological, climatological, anthropological, seismic, and land use/cover parameters and detailed landslide inventories are the foremost important input variables in GIS-based landslide susceptibility mapping. However, landslide frequency and magnitude are additional parameters in landslide hazard mapping. The susceptibility, hazard, and risk map produced from the expert judgment have a subjective problem for weight rating of the consequences of sets of parameters; however, the statistical analysis helps to develop maps supported the statistical relationship between sets of parameters and past or current landslide inventory data. Detailed landslide susceptibility, hazard, and risk map are often also developed for selected purpose at large scales using physical-based approaches. During this case, geotechnical properties of soil or rock slope material, angle of slope, and pore water pressure are the foremost important parameters to get a landslide susceptibility map supported the extent of an element of safety. Then, the hazard map are often produced by considering the factor of safety, landslide frequency, and magnitude. The danger map also can produce on large scale. Finally, the accuracy of the small-scale and detailed models are often validated using landslide inventory data using different techniques.

2.7 Landslide susceptibility and hazard mapping approaches

Landslide susceptibility or hazard zonation is a technique used to classify the slope into zones based on the level of actual or potential landslide susceptibility and hazard. Landslide susceptibility and hazard zonation are important for a rapid assessment of slope stability over a large area [21]. Landslide susceptibility map can forecast/provide important information about the spatial future landslide occurrence [3]. However, a landslide hazard map can forecast the spatial and temporal future landslide occurrence. In landslide susceptibility and hazard mapping, several approaches are developed, which may be categorized into qualitative, semi-quantitative, and quantitative methods [41–45].

2.7.1 Qualitative (expert evaluation) method

The expert evaluation method is a widely used technique, but a relatively subjective approach that explains the level of landslide condition in a descriptive expression based on the decision of the expert. Qualitative methods are an expert-driven approach, which required field experience specialists [41, 43, 45–49]. Field geomorphological analysis, landslide inventory analysis, and parameter assignment superimposition are the main activities for qualitative landslide susceptibility, and hazard mapping. Relying on the experience and professional background knowledge of experts and subjectivity is the drawback of these methods [41, 43, 45–47, 49]. This method has included heuristic, landslide inventory mapping, landslide hazard evaluation factor and slope stability evaluation parameter.

2.7.1.1 Heuristic method

This method is opinion based that is used to classify landslide susceptibility and hazard maps by mapping all landslide factors, and landslide through proper rating each factor classes to prepare a landslide susceptibility and hazard map. The demerits of this method are its subjectivity.

2.7.1.2 Landslide inventory method

Inventory is a simple method, which records the location and dimension of events occurred in the given area [50]. Landslide inventory is the way that used to record landslide location, size, occurrence time, displace material and types of slope failure. This method has used as the base for landslide susceptibility, hazard, and risk assessments; however, it does not provide the spatial relationship between landslide and sets of landslide factors rather than it only shows the location and volume of a landslide [51]. In this approach, landslide data can obtain through field mapping, historical record, satellite image or Google Earth Imagery analysis, and aerial photograph interpretation [36, 52].

2.7.1.3 Landslide Hazard evaluation factors (LHEF)

According to Anbalagan [21] this method is used for landslide susceptibility and hazard zonation /mapping with consideration of the inherent controlling factors only. It is simple and cost-effective over a large area. Nevertheless, this method has the following limitations.

   i. Has a rating of low value for groundwater effect on slope instability.

   ii. It does not account the triggering factors.

   iii. The condition of the rock mass with structural discontinuity and characteristics of the structural discontinuity (roughness, aperture, etc.) are not considered.

   iv. It is Subjective

   v. Give the same rating for lithology and structural discontinuity but discontinuities have great influence than lithology.

2.7.1.4 Slope stability evaluation parameters (SSEP)

Slope stability evaluation parameters (SSEP) is a landslide hazard zonation technique that is used to evaluate both inherent (slope material, slope geometry, structural discontinuity, land use and land cover, groundwater) and external factors (rainfall, seismicity, and human activity) to prepare landslide susceptibility map. Raghuvashi et al. [24], develop this method considering the dynamic and static landslide causative parameters. This technique is simple and supported by much field data but it is subjective for weighting assignment.

2.7.2 Semi quantitative method

Semi-quantitative methods are the combination of qualitative and quantitative methods, which introduce grading and weighting of the effects of landslide factors on landslide incidence [42, 53, 54]. In this method, both qualitative and quantitative methods can be applied to evaluate the effects of landslide governing factors on landslide occurrence [55]. Analytical hierarchy process, weighted linear combination, and expert knowledge/heuristic [42, 48, 56–59] are examples of semi-quantitative methods. Although some statistical concepts are introduced in this method, it depends on the expert's experience and the background of professional knowledge and some subjectivity remains [42, 60].

2.7.3 Quantitative (statistical) method

According to Canoglu [61]; Chen et al. [62], the quantitative methods are grouped into three categories such as machine learning/data mining, physical-based, and statistical methods. The statistical methods are indirect methods which is extensively or routinely used to assess the association between landslide governing factors and landslides based on mathematical [9, 41]. They are classified into multivariate and bivariate statistical methods [3]. The statistical methods are provided reliable results [4, 26, 42, 63–69]. The numerical methods rely on the mathematical model, expression, and less expert judgments, which provides comparatively reliable results, unlike the qualitative method. Among quantitative methods, the statistical method is the one, which used to evaluate the spatial slope instability based on the relationship between the past/active landslide and landslide factors [70]. A statistical method is an indirect method used to prepare a landslide hazard/susceptibility map, which is considered as objective and worked by integrated GIS tool with statistical analysis based on the landslide and sets of landslide factors spatial relationship. However, in this method, the most difficult thing that we have to consider is accurate database construction, model calibration, and model validation iteration procedures [71]. In this method, each factor has mapped and overlaid over past/active landslides to carry out the contribution of each factor and subclass on the instability of the slope [24, 52, 72]. The limitation of the statistical method is its requirement for detailed and quality landslide and landslide factor data, and it is time-consuming to acquire them over a large area Raghuvashi et al. [24]. The statistical method cannot apply to the area where a landslide has not occurred. This is one of the limitations of statistical methods in landslide susceptibility, hazard, and risk mapping.

2.7.3.1 Bivariate statistical analysis

The bivariate statistical procedure is straightforward to use and update, which is capable to differentiate the consequences of every sub factor class for landslide occurrence. Within the bivariate statistical procedure, the presence of landslide has been considered because the variable and therefore the parameters that enhanced the occurrence of the landslide has been considered as the independent variable [73]. In this technique, each determinant map has been classified into sub-classes to work out the response of individual factor classes to landslide occurrence. The landslide factor classes are often combined with a landslide distribution map and weighting values supported the landslide densities of every determinant class. After weight value calculation, the weighted raster map is carefully sum up employing a raster calculator in Math algebra under the GIS tool to urge the landslide susceptibility index map. The landslide susceptibility or hazard index map are often reclassified using various methods like natural break under the GIS tool to urge the ultimate landslide susceptibility map. The benefits of bivariate statistical methods are they will cover an outsized area with effective cost; it is simple to apply; it can provide spatially distributed landslide information and its relationship with landslide factors. However, the bivariate statistical methods have the subsequent limitation 1. It cannot distinguish which factor is more influential and non-influential. 2. It cannot provides the knowledge about the inherent condition of the slope material like geotechnical method 3. It can predict the landslide susceptibility regions but it cannot be predicted when this landslide will occur and it needs landslide occurrence during a certain region to predict the opposite region which has some environmental factor. The load of evidence, information value, certainty factor, and frequency ratio is that the commonest techniques in bivariate statistical analysis.

2.7.3.2 Multivariate statistical analysis

This method will provide more realistic and accurate results. It also considers the mutual relationship among landslide factors, unlike bivariate statistical methods. The weight of causal factors indicates the relative contribution of every factor to the degree of hazard in a given land unit. The multivariate statistical procedure helps to perform multivariate statistical analysis unlike the bivariate statistical procedure. One among the merits of the multivariate method is capable to work out the influential power of individual landslide factors on landslide occurrence. Logistic regression, discriminant analysis, and cluster analysis are the foremost commonly applied techniques in this method.

2.7.3.3 Data mining method

In recent times, advanced data mining methods have been widely used in landslide susceptibility modeling., including random forest [56–58], boosted regression tree [74], classification and regression tree [74], Naïve Bayes [53, 75], support vector machines [32, 76], kernel LR [77], logistic model tree [56–58, 77], index of entropy [39], and artificial neural networks [56–58, 78, 79]. Data mining methods are incapable to work out the consequences of every landslide factor class, need high computing capacity, time-consuming, and therefore the internal calculation process of those methods is intensive and cannot easily be understood. Although both statistical and data mining methods have a bit little difference in the degree of predictive accuracy, they can provide reliable predictive accuracy landslide susceptibility map in landslide susceptibility or hazard mapping [78, 80].

2.7.3.4 Physical based approach

The physical-based approach includes limit equilibrium and finite element numerical models. These methods can be applied for both soil and rock slope stability analysis. This method can provide hazard in absolute value /factor of safety or probability/quantitative results that can directly use for design purposes [52] and Raghuvashi et al. [24]. Physical-based methods are used to calculate the quantitative value of the inherent slope materials of the factor of safety over a defined area [81]. These methods can be applied when landslide types are simple (shallow landslides) and the intrinsic properties of slope material are homogeneous [81]. It requires detailed ground data such as unit weight of soil, soil strength, soil layer thickness, slope angle, pore water pressure, depth below the terrain surface, and slope height. The physical-based method has been employed over a small area, and oversimplification, data availability to acquire frequently is impossible are the drawback of these methods [81]. These methods can be focused on an on-site investigation to assess the geotechnical properties of soil/rock, soil depth, surface and subsurface water condition, the geometry of the slope, landslide location, failure mechanism, depth, and distance of landside. These methods are used to analyze slope conditions by calculating factors of safety using different software like PLAXIS and Slope/w in GeoStudio software package as two or three-dimensional models. The oversimplification of geological, geotechnical model and the difficulty to predict pore water pressure and its relationship with rainfall /snowmelt are the main problems that challenge use the of geotechnical approaches [82].

2.8 Landslide risk mapping approaches

Landslide risk is the expected loss or damage due to landslide Incidences, which include fatalities, damage to properties, infrastructure, farmland, environment, interruption of services, and economic activities. As compare to landslide susceptibility, and hazard mapping, landslide risk mapping is not common so far due to it requires complex input parameters. It is a complex task due to the lack of necessary information to produce input parameters including vulnerability/susceptibility, hazard, and element at risk [33]. In addition to landslide susceptibility/vulnerability, and hazard maps, landslide risk map is very important in the regulation of land use, landslide risk management, and mitigation strategies. One has a plan to prepare a landslide risk map, it is necessary to estimate landslide susceptibility, hazard, and element at risk.

In landslide risk mapping, qualitative and quantitative techniques are commonly practiced methods. The qualitative (heuristic) method is used to estimate the level of risk in an area qualitatively, when the numerical estimation of hazard, vulnerability, and element at risk is difficult due to lack of landslide frequency, date of occurrence, and magnitude data [33, 83]. The landslide risk map can be produced based on the knowledge of experts about landslide vulnerability, hazard, and element at risk. In a quantitative approach, landslide risk can be estimated numerically using a mathematical equation developed by Varnes and IAEG Commission on landslides and other mass movements on slopes (1984). Risk = hazard*vulnerability*element at risk. Where the hazard is the probability of landslide occurrence in a particular type and magnitude in a given location within a referenced period. Vulnerability is the expected degree of loss due to landslides. Element at risk is potentially affected elements in landslide-affected areas.

2.9 Landslide susceptibility, hazard, and risk model validation

In the case of model validation, landslide area has been classified based on time, space, and random partition [26, 84, 85]. The model can be validated by applied various validation techniques like predictive rate curve, success rate curve, simple overlay, a landslide percent comparison column chart, relative error, relative landslide density index (R – index), receiver operating characteristics (ROC), and landslide density.

2.9.1 Success and predictive rate curve

As indicated in [26], the success rate curve can be plotted using training landslide against the landslide susceptibility or hazard or risk map area. Success rate and a predicted rate curve can be plotted using a cumulative percentage of training/testing landslide area against the cumulative percentage of the landslide susceptibility/hazard/risk map area [86]. For this purpose, the landslide susceptibility or hazard or risk index has to be reclassified into 100 classes by descending order of the value. Then landslide raster can be combined with these classes to obtain landslide pixels. Both landslide and map area pixels have converted into a cumulative percentage to plot the success and predicted rate curve. The success rate curve can be plotted using the cumulative percentage of training landslide vs. a cumulative percentage of map area while the predicted rate curve can be plotted using a cumulative percentage of testing or validation landslide area vs. map area. The success rate explains how well the model and how landslide susceptibility, hazard, and risk mapping results are classified the study area using training landslide data. The predict rate curve explains the predictive capability of the conditioning factor for the model. If the curve deflects and closes to the top left of the reference line along the diagonal, the model has higher accuracy.

2.9.2 Landslide density

As states by Pham et al., [87] and Fayez et al., [88], the landslide density has calculated using the equation of landslide density (LD). $LD = \dfrac{\text{percent of observed landslide}}{\text{percent of predicted landslide}}$. The higher landslide density on the high, and very high landslide susceptibility, hazard, and risk region confirms that the model is reliable and accurate [87].

2.9.3 Relative landslide density index (R: Index)

Landslide susceptibility, hazard, and risk models can also be validated using the relative landslide density index, which is calculated using the following equation. $R\text{-}Index = \dfrac{\dfrac{ni}{Ni}}{\sum \dfrac{ni}{NI} * 100}$. Where ni is the number of landslide in a landslide susceptibility classes while Ni is the number of landslide susceptibility/hazard/risk class pixel within that class. The relative density can calculate using an equation through a comparison of landslide susceptibility with landslide inventory data set [73, 89]. As the R- index value increases from the very low to very high landslide susceptibility/hazard/risk classes confirms that the model is accurate and reliable.

2.9.4 Relative error

The other model validation technique relative error calculation is one of the techniques that help us to evaluate and determine the quality of the model and the number of landslides in the higher landslide susceptibility, hazard/risk classes.

The higher the relative error value the poorer the model accuracy. When the relative error greater than 0.5, the model is not acceptable [90]. However when the relative error less than or equal to 0.5 and the number of landslide in the high landslide susceptibility/hazard/risk class more than half, the given model is accurate and reliable. *Relative error* $(\xi) = \sum \text{TNLS} - \sum \text{NLS} / \sum \text{TNLS}$. Where TNLS is the total number of landslide in a region and NLS is a number of landslide in the high and very high landslide susceptibility/hazard/risk classes.

2.9.5 Receiver operating characteristics (ROC)

The ROC is the curve used to evaluate the performance of the landslide susceptibility, hazard, and risk models. ROC curve is the graphical representation of true positive rate (TPR) as y-axis and false positive rate (FPR) as x-axis. In the ROC curve, the area under the curve (AUC) is the most important diagnostic feature that helps to evaluate whether the model performance is accurate or not accurate. As stated by Yesilnacar and Topal [91], the value of AUC is usually found in between 0.5–1. The model has excellent performance when the AUC value is in between 0.9–1; the model has very good performance when the AUC value is in between 0.8–0.9. The model has good performance when the AUC value is between 0.7–0.8. When the value of AUC is between 0.6–0.7, the model has average performance however if the AUC value is between the range of 0.5–0.6 and equal to 0.5 or less than o.5, the model has poor and useless results.

2.10 Case study on landslide susceptibility mapping

2.10.1 Landslide susceptibility mapping using statistical methods in Uatzau catchment area, Northwestern Ethiopia

Recent unconsolidated soil deposits, rugged topography, active gully, riverbank erosion, and improper land use practice characterize the study area (Uatzau), making it vulnerable to a variety of landslides, including earth fall, soil creep, weathered rockslide, soil slide, earth flow, and debris flow. Landslide susceptibility zones of the study area were determine using Frequency ratio (FR), certainty factor (CF), and information value (IV) models. These maps also depict the spatial distribution of projected landslides and the locations where they are expected to occur. The maps, on the other hand, may not be able to predict the amount of material that will be displaced, as well as the time and frequency with which the landslide will occur. The landslide susceptibility models can also helpful for preventative and mitigation measure of landslide hazard in regional land use planning [81, 82, 92–96]. The success rate curve and predictive rate curve were used to validate the maps using training and testing/validation landslide data sets. The success rate curve was used to assess how successfully the models identified the location and supported the landslide events that were occurring at the time [26, 96]. The prediction rate curve was created to assess how effectively the models can forecast future landslide events that are unknown [94, 96]. Within the region, steep slopes covered by very loose shallow soil deposits, closer to the stream, agricultural land on a steep slope, active gully erosion, and concave slope shapes resulted in the high and very high susceptibility classes, while the moderate susceptibility class is found in highland landscapes. Low plain landscapes and areas covered by vast weathering-resistant rock masses are into the realm of very low and low susceptibility of a region.

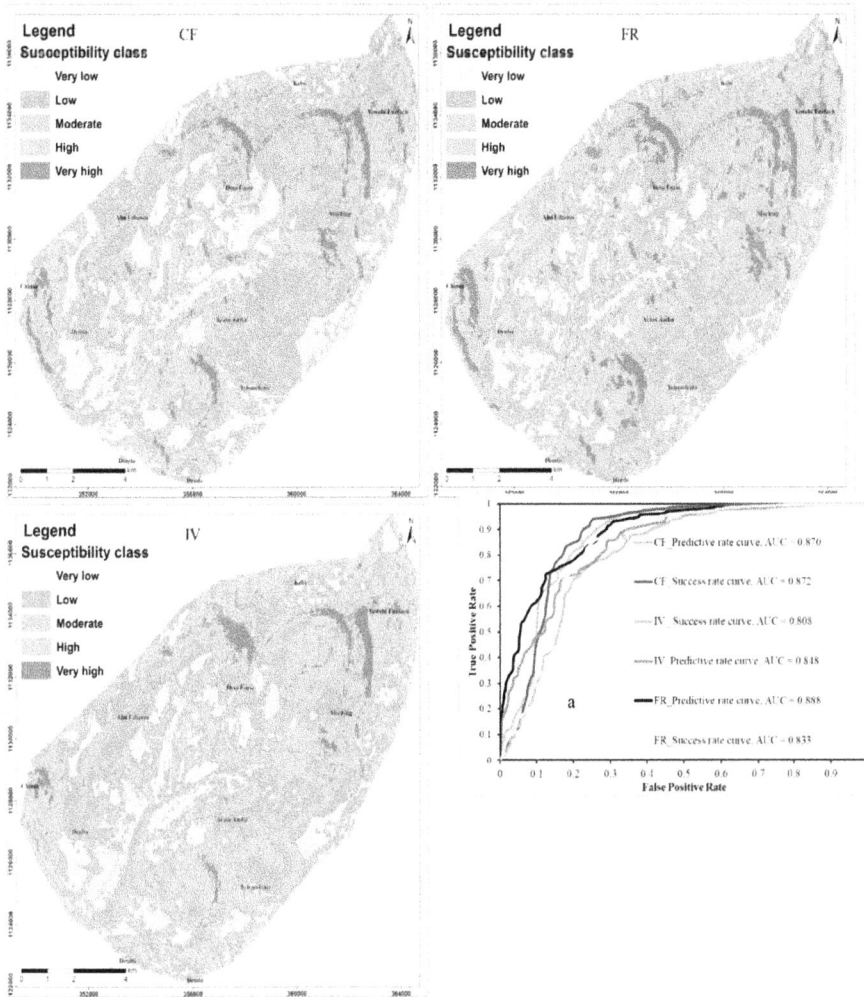

Figure 5.
Landslide susceptibility maps of frequency ratio (FR), certainty factor (CF), information value (IV) methods [36] and a) receiver operating characteristics curve (ROC) [36].

Zine et al. [97] stated that higher prediction accuracy (AUC = 89.05%) and AUC = 85.57%) was received using the information value and frequency ratio methods. Similarly, the frequency ratio approach outperformed the information value methods for both success rates (AUC = 83.27%) and prediction rate curve (AUC = 88.8%) in this investigation. The accuracy of the two models falls within the same ranges, which may be a good performance. The frequency ratio model revealed a slight difference in the AUC value. Qiqing et al. [40] stated that a high predictive accuracy of AUC value of 75 was received using a certainty factor model when compared to the prediction rate curve value (AUC = 64.08%) of information value model. However, their accuracy values were within the same ranges, suggesting that they performed well. Similarly, in the current model, the certainty factor model had a greater prediction rate value (AUC = 87.03%) than the information value model, which had a lower prediction rate value (AUC = 84.8%), but they both required an equivalent accuracy range, which may be a good performance. The work of Haoyuan et al. [98] supported the predictive rate value of the area

Information value method	LSI Value	LSI	Factor class area (%)	Validation data set (%)	Training data set (%)	AUC for validation landslide	AUC for training landslide
	−0.5-0.9	VLS	15.5	3.9	6.3	0.848323	0.808265
	0.9–1.5	LS	24.3	7.9	11.8		
	1.5–2	MS	31.5	20.1	23.8		
	2.0–2.6	HS	21.1	40.8	31.8		
	2.6–4.1	VHS	7.6	27.3	26.3		
Certainty Factor (CF)	−2.2- −0.97	VLS	17.8	4.7	6.0	0.870348	0.871933
	-0.97- −0.47	LS	31.0	12.3	16.4		
	-0.47-0.04	MS	28.8	17.5	24.8		
	0.04–0.74	HS	19.0	34.8	33.0		
	0.74–2.61	VHS	3.4	30.7	19.7		
Frequency Ratio (FR)	3.1–4.3	VLS	22.7	5.4	9.3	0.888337	0.832718
	4.3–4.8	LS	30.8	14.7	17.8		
	4.8–5.3	MS	22.4	19.5	20.0		
	5.3–6	HS	19.3	43.7	35.0		
	6–7.7	VHS	4.8	16.7	17.8		

VLS is for very low susceptibility, LS stands for low susceptibility, MS stands for moderate susceptibility, HS stands for high susceptibility, VHS stands for very high susceptibility, LSI stands for landslide susceptibility index and AUC stands for area under the curve.

Table 3.
Statistical summary of information value, certainty factor, and frequency ratio methods [36].

under the receiver operating characteristic curve (AUC), showing that the frequency ratio and certainty factor models have the more or less similar predictive capacity, with the certainty factor model having 81.18% and the frequency ratio model having 80.14%, respectively. The Frequency ratio model, on the other hand, performed worse than the CF model. The two models in this study had essentially identical AUC values for the prediction rate curve (87.03% for the certainty factor model and 88.8% for the frequency ratio model) (**Figure 5**). The closer prediction capacity with AUC > 64% and AUC > 80%, respectively, fall within the range of good and extremely good performance, according to the three bivariate statistical methods in the literature and this work [91]. High and extremely high susceptibility classes encompassed nearly 20% of the research area in this study (**Table 3**). The landslide validation findings for the three models are more similar than they are dissimilar, and they are all in the same region of outstanding performance. Aside from that, the percentages of landslides that fall into the high and highly susceptible classes are nearly the same (60.4%, 65.5% & 68.1% for FR, CF, and IV, respectively). Because of these findings, the research effort concludes that in landslide susceptibility mapping, the three models have similar potential for identifying landslide-prone locations, although factor selection should take precedence over methodologies. However, when compared to the FR and CF approaches, the IV models' moderate, high, and very high susceptibility area coverage exhibited minor differences in a single example. This is frequently due to flaws discovered in IV during weight rating for each factor class, i.e. when there is no landslide in a certain component class the IV results become zero. This gives a good indication of the model's overall accuracy. FR and CF models are better for regional land use planning, landslide hazard mitigation, and prevention based on the prediction accuracy of AUC value. Although the generated maps cannot predict when and how often landslides will occur, they do show the spatial distribution of land-slide risk.

2.11 Conclusion

This chapter introduces and overview the concepts of landslide, type, factors, inventories, susceptibility, hazard, and risk. Moreover, different mapping and validation approaches were introduced. The confusing between the term suscepti-bility and hazard is clearly discussed. Detail and quality data should tend emphasis in getting quality landslide susceptibility, hazard, and risk maps. Field landslide investigation integrated with Google Earth Imagery analysis is vital to work out and record, the relative occurrence date, magnitude, dimension, type, and state of landslide. GIS-based landslide susceptibility, hazard, and risk mapping is suitable for regional scale where as physical based mapping is recommendable for detail landslide study where geotechnical investigation is require.

Acknowledgements

First and foremost, I want to express my gratitude to the almighty God for allow-ing me to complete this study project. Next, I would like to express my gratitude to my wonderful family and friends for their unwavering support during the research process. Finally, I would like to express my gratitude to the University of Gondar.

Contributions of the authors

I was responsible for all aspects of the project, including the conception and design of the work, model development, statistical analysis, and interpretation of the results.

Funding

In this scenario, it is not appropriate.
Data and materials are readily available.
The corresponding author can provide all of the datasets that were utilized and analyzed during the current investigation.

Interests in computing

There are no competing interests.

Author details

Azemeraw Wubalem
Department of Geology, University of Gondar, Gondar, Ethiopia

*Address all correspondence to: alubelw@gmail.com

IntechOpen

References

[1] Hervas J. Leson learnt from landslide disasters in Europ. JPC report EUR 20558 EN. Office for official publications of European communities Luxemburg. 2003:91

[2] Hong H, Miao Y, Liu J, Zhu AX. Exploring the effects of the design and quantity of absence data on the performance of random forest-based landslide susceptibility mapping. CATENA. 2019;**176**:45-64

[3] Azemeraw W. Modeling of Landslide susceptibility in a part of Abay Basin, northwestern Ethiopia. Open Geosciences. 2020;**12**:1440-1467. https://doi.org/10.1515/geo-2020-0206

[4] Azemeraw W, Meten MAG. Landslide susceptibility mapping using information value and logistic regression models in Goncha Siso Eneses area, northwestern Ethiopia. SN Applied Sciences, Switzerland. 2020;**2**:807 https://doi.org/10.1007/s42452-020-2563-0

[5] Aleotti P, Chowdhury R. Landslide hazard assessment: summary review and new perspectives. Bulletin of Engineering Geology and the Environment. 1999;**58**:21-44

[6] Gutiérrez F, Linares R, Roqué C, Zarroca M, Carbonel D, Rosell J, et al. Large landslides associated with a diapiric fold in Canelles reservoir (Spanish Pyrenees): detailed geological–geomorphological mapping, trenching and electrical resistivity imaging. Geomorphology. 2015;**241**:224-242

[7] Jazouli, El A., Barakat, A. & Khellouk, R. GIS-multicriteria evaluation using AHP for landslide susceptibility mapping in Oum Er Rbia high basin (Morocco). *Geoenviron Disasters*. 2019;**6**:3. https://doi.org/10.1186/s40677-019-0119-7

[8] Rai PK, Mohan K, Kumra VK. Landslide hazard and its mapping using remote sensing and GIS. Journal of Scientific Research. 2014;**58**:1-13

[9] Chen Z, Wang J. Landslide hazard mapping using a logistic regression model in Mackenzie Valley. Canada. Nat. Hazard. 2007;**42**(1):75-89

[10] Brabb E. Innovative Approaches for Landslide Hazard Evaluation. Toronto: IV International Symposium on Landslides; 1984. pp. 307-323

[11] Kifle Woldearegay. Review of the occurrences and influencing factors of landslides in the highlands of Ethiopia with implications for infrastructural development. Mekelle University, Mekelle, Ethiopia. Journal of Muna. 2013b;5:331

[12] Australian Geomechnics Society Landslide zoning Working Group. Guidline for landslide susceptibility, hazard and risk zoning for landuse planning. Australian Geomechnics Society. 2007;**42**:1-27

[13] Varnes DJ. Slope movement types and processes. In: Schuster RL, Krizek RJ, editors. Landslides, analysis and control, special report 176: Transportation research board. Washington, DC: National Academy of Sciences; 1978. pp. 11-33

[14] U.S. Geological Survey. Landslide Types and Processes. 2004

[15] Msilimba G. A comparative study of landslides and geohazard mitigation in Northern and. Central Malawi; 2007

[16] Washington Geological Survey (WGS). What are landslides and how do they occur? 2017;

[17] Keefer DK. Statistical analysis of an earthquake-induced landslide

distribution the 1989 Loma Prieta, California event. Eng. Geol. 2000;**58**:231-249

[18] Gorsevski P.V., Jankowski, P., and Gessler, P.E. A heuristic approach for mapping landslide hazard by integrating fuzzy logic with the analytic hierarchy process. Control and Cybernetics. 2006;**35**(1):121-146

[19] International geotechnical societies UNESCO Working party on landslide inventory (WP/WLI). Suggested method for describing the cause of landslide. Bull Intern Assoc Eng Geol. 1994;**50**:71-74

[20] Vallejo LI, González de, Ferrer Mercedes. Geological engineering. Taylor & Francis. Group. 2011;**692**

[21] Anbalagan R. Landslide hazard evaluation and zonation mapping in mountainous terrain. Eng. Geol. 1992;**32**:269-277

[22] Ayalew L, Yamagishi H. Slope failures in the Blue Nile basin, as seen from landscape evolution perspective. Geomorphology. 2004;**57**:95-116

[23] Hamza T, Raghuvanshi TK. GIS-based landslide hazard evaluation and zonation in Jeldu district in central Ethiopia. Journal of king saud university science. 2016;**29**:151-165

[24] Raghuvanshi TK, Ibrahim J, Ayalew D. Slope Stability Susceptibility evaluation parameter (SSEP) rating scheme: An approach for landslide hazard zonation. Journal of African Earth Sciences. 2015;**99**:595-612

[25] Lai R. Effects of slope length on runoff from Alfisols in Western Nigeria. Geoderma. 1983;**31**:185

[26] Meten M, Bhandary NP, Yatabe R. GIS-based frequency ratio and logistic regression modeling for landslide susceptibility mapping of Debre Sina area in central Ethiopia. J. Mt. Sci. 2015;**12**(6):1355-1372

[27] Xu C, Dai F, Xu X, Lee YH. GIS – based support vector machine modeling of earthquake – triggered landslide susceptibility in the Jianjiang river watershed, China. Geomophology. 2012;**145**-146:70-80

[28] Pradhan B, Lee S, Buchroithner MF. Remote sensing and GIS-based landslide susceptibility analysis and its cross-validation in three test areas using a frequency ratio model. Photogramm Fernerkun. 2010;**1**:17-32. DOI: 10.1127/14328364/2010/0037

[29] Pradhan B, Lee S. Landslide susceptibility assessment and factor effect analysis: backpropagation artificial neural networks and their comparison with frequency ratio and bivariate logistic regression modelling. Environmental Modelling & Software. 2010;**25**:747-759

[30] Alkhasawneh MS, Ngah UK, Tay LT, et al. Determination of important topographic factors for landslide mapping analysis using MLP Network. Hindawi Publishing Corporation. The Scientific World Journal. Article ID. 2013;**415023**

[31] Abebe B, Dramis F, Fubelli G, Umer M, Asrat A. Landslides in the Ethiopian highlands and the Rift margins. Journal of African Earth Sciences. 2010;**56**:131-138

[32] Lin GF, Chang MJ, Huang YC, Ho JY. Assessment of susceptibility to rainfall-induced landslides using improved self-organizing linear output map, support vector machine, and logistic regression. Eng Geol. 2017;**224**:62-74

[33] Guzzetti F, Reichenbach P, Cardinali M, Galli M, Ardizzone F. Landslide hazard assessment in the Staffora basin, northern Italian Apennines. Geomorphology. 2005

[34] Mohammad M, Pourghasemi HR, Pradhan B. Landslide susceptibility mapping at Golestan Province, Iran: a comparison between frequency ratio, Dempster–Shafer, and weights-of evidence models. J Asian Earth Sci. 2012;**61**:22136

[35] Guzzetti F, Cardinali M, Reichenbach P, Cipolla F, Sebastiani C, Galli M, et al. Landslides triggered by the 23 November 2000 rainfall event in the ImperiaProvince, Western Liguria. Italy. Engineering Geology. 2004;**73**(2):229-245

[36] Azemeraw W. Landslide susceptibility mapping using statistical methods in Uatzau catchment area, northwestern Ethiopia. Geoenvironmental Disasters. 2021;**8**(1):1-21. https://doi.org/10.1186/s40677-020-00170

[37] Guzzetti F, Carrara A, Cardinal M, Reichenbach P. Landslide hazard evaluation: a review of current techniques and their application in a multi-scale study, central Italy. Geomorphology. 1999;**31**(1-4):181-216

[38] Saha AK, Gupta RP, Sarkar I, Arora KM, Csaplovics E. An approach for GIS-based statistical landslide susceptibility zonation with a case study in the Himalayas. Landslides. 2005;**2**(1):61-69

[39] Hong H, Chen W, Xu C, Youssef AM, Pradhan B, Bui DT. Rainfall-induced landslide susceptibility assessment at the Chongren area (China) using frequency ratio, certainty factor, and index of entropy. Geocarto Int. 2016; https:// doi.org/10.1080/10106049.2015.1130086

[40] Wang Q, Guo Y, Li W, He J, Wu Z. Predictive modeling of landslide hazards in Wen County, northwestern China based on information value, weights-of-evidence, and certainty factor, Geomatics. Natural Hazards and Risk. 2019;**10**(1):820-835. DOI: 10.1080/19475705.2018.1549111

[41] Bednarik M, Yilmaz I, Marschalko M. Landslide hazard and risk assessment: a case study from the Hlohovec–Sered' landslide area in south-west Slovakia. Nat Hazards. 2012. DOI: 10.1007/s11069-012-0257-7

[42] Hong H, Junzhi L, A-Xing Z. Modeling landslide susceptibility using logitBoost alternating decision trees and forest by penalizing attributes with the bagging ensemble. Science of the Total Environment. 2020;**718**:3-15

[43] Pradhan B, Mansor S, Pirasteh S, Buchroithner M. Landslide hazard and risk analyses at a landslide-prone catchment area using the statistical-based geospatial model. Int J Remote Sens. 2011a;**32**(14):4075-4087. DOI: 10.1080/01431161.2010.484433

[44] Regmi AD, Yoshida K, Pourghasemi HR, Dhital MR, Pradhan B. Landslide susceptibility mapping along Bhalubang-Shiwapur area of mid-western Nepal using frequency ratio and conditional probability models. Jour. Mountain Sci. 2014;**11**(5):1266-1285

[45] Wang HB, Wu SR, Shi JS, Li B. Qualitative hazard and risk assessment of landslides: a practical framework for a case study in China. Nat Hazards. 2011. DOI: 10.1007/s11069-011-0008-1

[46] Jia N, Xie M, Mitani Y, Ikemi H, Djamaluddin I. A GIS-based spatial data processing system for slope monitoring. Int Geoinf Res Dev J. 2010;**1**(4)

[47] Varnes DJ. Landslide hazard zonation, a review of principles and practice, International Association of Engineering Geology Commission on Landslides and Other Mass Movements on Slopes, UNESCO. Paris. 1984;**63**

[48] Wang Y, Fang Z, Mao W, Peng L, Hong H. Comparative study of landslide

susceptibility mapping with different recurrent neural networks. Computers and Geosciences. 2020;**138**:10445

[49] Karimi Nasab S, Ranjbar H, Akbar S. Susceptibility assessment of the terrain for slope failure using remote sensing and GIS, a case study of Maskoon area. Iran. Int Geoinf Res Dev J. 2010;**1**(3)

[50] Ayenew T, Barbieri G. Inventory of Landslides and Susceptibility Mapping in the Dessie area, Northern Ethiopia. Elsevier, Engineering Geology. 2005;**77**:1-15

[51] Casagli N, Catani F, Puglisi C, Delmonaco G, Ermini L, Margottini C. An Inventory-based approach to landslide susceptibility assessment and its application to the Virginio River Basin. Italy. Environ. Eng. Geosci. 2004;**3**:203-216

[52] Paradeshi, S.D., Sumant, E. Atade, and Suchitra, Paradeshi, S. landslide hazard assessment: recent trends and techniques. Springer open journal. 2013; 2: 1-11

[53] Pham BT, Bui DT, Pourghasemi HR, Indra P, Dholakia MB. Landslide susceptibility assessment in the Uttarakhand area (India) using GIS: a comparison study of prediction capability of nave bayes, multilayer perceptron neural networks, and functional trees methods. Theor Appl Climatol. 2017;**128**:255-273

[54] Tie Bui D, Shahabi H, Geertsema M, Omidvar E, Clagu J. J, Thai Pham B, Dou J, Talebpour ASLD, Bin Ahmad B, Lee S. New ensemble models for shallow landslide susceptibility modeling in a semi arid watershed. Forests. 2019;**10**(9):743.

[55] Tie Bui D, Shahabi H, Omidvar E, Shizardi A, Geertsema M, Clagu J. J, Khosrovi K, Pradhan B, Pham B. T, Chapi K, Barati Z. (2019). Shallow

landslide prediction using a novel hybrid functional machin learing algorthism. Remote Sens. 11(8):931.

[56] Chen W, Pourghasemi HR, Kornejady A, Zhang N. Landslide spatial modeling: introducing new ensembles of ANN, MaxEnt, and SVM machine learning techniques. Geoderma. 2017;**305**:314-327

[57] Chen W, Pourghasemi HR, Zhao Z. A GIS-based comparative study of Dempster-Shafer, logistic regression and artificial neural network models for landslide susceptibility mapping. Geocarto Int. 2017;**32**:367-385

[58] Chen W, Xie X, Wang J, Pradhan B, Hong H, Bui DT. A comparative study of the logistic model tree, random forest, and classification and regression tree models for spatial prediction of landslide susceptibility. CATENA. 2017;**151**:147-160

[59] Zhu AX, Miao Y, Wang R, Zhu T, Deng Y, Liu J, et al. A comparative study of an expert knowledge-based model and two data-driven models for landslide susceptibility mapping. CATENA. 2018;**166**:317-327

[60] Tsegaratos P, Ilia I, Hong H, Chen W, Xu C. Applying information theory and GIS based quantitative methods to produce landslide susceptibility maps in mancheang county. China. Landslides. 2017;**14**:1091-1111

[61] Canoglu MC. Deterministic landslide susceptibility assessment with the use of a new index (factor of safety index) under dynamic soil saturation: an example from Demircikoy watershed (Sinop/Turky). Carpathian journal of Earth and Environmental Sciences. 2017;**12**:423-436

[62] Chen et al. (2019). Spatial prediction of landslide susceptibility using data mining based kernel logistic

regression, naive Bayes, and RBFNetwork for long county area (China). Bull. Eng. Geol. Environ. 247-266.

[63] Luelseged A, Yamagishi H. The application of GIS-based logistic regression for landslide susceptibility mapping in the Kakuda- Yahiko Mountains, Central Japan. Geomorphology. 2005;**65**:15-31

[64] Chandak, P.G., Sayyed, S.S., Kulkarni, Y.U., Devtale, M. K (2016) Landslide hazard zonation mapping using information value method near Parphi village in Garhwal Himalaya. Ljemas, 4: 228 – 236

[65] Dai FC, Lee CF. Landslide characteristics and slope instability modeling using GIS, Lantau Island, Hong Kong. Geomorphology. 2002;**42**:213-228

[66] Donati, L, and Turrini, M. C. (2002). An objective method to rank the importance of the factors predisposing to landslides with the GIS methodology application to an area of the Apennines (Valnerina; Perugia, Italy). Engg. Geol. 63: 277-289.

[67] Duman, T.Y., Can, T., Gokceoglu, C., Nefesliogocu, H. A., and Sonmez, H. (2006). Application of logistic regression for landslide susceptibility zoning of Cekmee area, Istanbul, Turkey. Verlag. 242 – 256.

[68] Kouhpeima S. Feizniab H. Ahmadib and Moghadamniab A.R. (2017). Landslide susceptibility mapping using logistic regression analysis in Latyan catchment. Desert. 85 – 95.

[69] Sarkar S, Rjan MT, Roy A. Landslide susceptibility Assessment using information value method in parts of the Darjeeling Himalayas. Geological Society of India. 2013;**82**:351-362

[70] Carrara A, Cardinali M, Guzzetti F. Uncertainty in assessing landslide hazard and risk. ITC Journal. 1992;**2**:172-183

[71] Ercanoglu M, Gokceoglu C, Van Asch TWJ. Landslide susceptibility zoning of North of Yenice (NW Turkey) by muti-variate statistical techniques. Nat. Haz. 2004;**32**:1-23

[72] Girma F, Raghuvanshi TK, Ayanew T, Hailemariam T. Landslide hazard zonation in Ada Berga district, Central Ethiopia, A GIS-based statistical Approach. Journal of Geomatics. 2015;**9**:1-14

[73] Shahabi BB, Khezri AS. Evaluation and comparison of bivariate and multivariate statistical methods for landslide susceptibility mapping (case study: Zab basin). Arab J Geosci. 2013;**6**:3885-3907

[74] Youssef AM, Pourghasemi HR, Pourtaghi ZS, Al-Katheeri MM. Landslide susceptibility mapping using random forest, boosted regression tree, classification and regression tree, and general linear models and comparison of their performance at Wadi Tayyah Basin, Asir Region, Saudi Arabia. Landslides. 2016;**13**:839-856

[75] Tsangaratos P, Ilia I. Comparison of a logistic regression and Naïve Bayes classifier in landslide susceptibility assessments: the influence of models complexity and training dataset size. CATENA. 2016;**145**:164-179

[76] Hong HY, Pradhan B, Xu C, Tien BD. Spatial prediction of landslide hazard at the Yihuang area (China) using two-class kernel logistic regression, alternating decision tree, and support vector machines. CATENA. 2015;**133**:266-281

[77] Bui DT, Tuan TA, Klempe H, Pradhan B, Revhaug I. Spatial prediction models for shallow landslide hazards: a comparative assessment of the efficacy of support vector machines, artificial

neural networks, kernel logistic regression, and logistic model tree. Landslides. 2016;**13**:361-378

[78] Soyoung P, Choi C, Kim B, Kim J. Landslide susceptibility mapping using frequency ratio, analytic hierarchy process, logistic regression, and artificial neural network methods at the Inje area. Korea. Environ Earth Sci. 2013;**68**:1443-1464

[79] Yilmaz I. Comparison of landslide susceptibility mapping methodologies for Koyulhisar, Turkey: conditional probability, logistic regression, artificial neural networks, and support vector machine. Environ Earth Sci. 2010;**61**:821-836

[80] Goetz JN, Brenning A, Petschko H, Leopold P. Evaluating machine learning and statistical prediction techniques for landslide susceptibility modeling. Comput Geosci. 2015;**81**:1-11

[81] Yilmaz I, Keskin I. GIS-based statistical and physical approaches to landslide susceptibility mapping (Sebinkarahisar, Turkey). Bull Eng Geol Environ. 2009;**68**:459-471

[82] Fell R, Corominas J, Bonnard C, Cascini L, Leroi E, Savage WZ. Guidelines for landslide susceptibility, hazard, and risk zoning for land-use planning, joint technical committee (JTC-1) on landslides and engineered slopes. Eng Geol. 2008;**102**:85-98

[83] Hervas Javier and Bobrowsky Peter. (2009). Mapping inventories, susceptibility, hazard, and risk.

[84] Chung CJF, Fabbri AG. Validation of Spatial Prediction Models for Landslide Hazard Mapping. Natural Hazards. 2003;**30**(3):451-472

[85] Lee S, Pradhan B. Landslide hazard mapping at Selangor. Malaysia using frequency ratio and logistic regression models, Landslides. 2007;**4**:33-41

[86] Omar Althuwayee, Pradhan, B., Mahmud, A. R. Prediction of slope failures using the bivariate statistical based index of entropy model. 2012:1-7

[87] Pham BT, Tien Bui D, Prakash I, Dholakia M. Landslide susceptibility assessment at a part of Uttarakhand Himalaya, India using GIS-based statistical approach of frequency ratio method. International Journal of Engineering Research and Technology. 2015;**4**:338-344

[88] Fayez L, Pazhman D, Binh TP, Dholakia MB, Solank HA, Khalid M. Application of frequency ratio model for the development of landslide susceptibility Mapping at Part of Uttarakhand State. India. International Journal of Applied Engineering. 2018;**13**(9):6846-6854

[89] Meena SR, Ghorbanzadeh O, Blaschke T. Validation of spatial prediction models for landslide susceptibility mapping by considering structural similarity. ISPRS Int J Geo Inform. 2019;**8**:94

[90] Fressard, M., Thiery, Y., and Maquaire, O. Which data for quantitative landslide susceptibility mapping at an operational scale? Case study of Paysd'Auge plateau hillslopes Normandy, France). Nat. Hazards Earth Syst.sci. 2014;**14**:569-588

[91] Yesilnacar E, Topal T. Landslide susceptibility mapping: A comparison of logistic regression and neural networks method in a medium scale study, Hendek region (Turkey). Engineering Geology. 2005;**79**:251-266

[92] Das G, Lepcha K. Application of logistic regression (LR) and frequency ratio (FR) models for landslide susceptibility mapping in Relli Khola river basin of Darjeeling Himalaya. India. SN Appl Sci. 2019;**1**:1453. https://doi.org/10.1007/s4245 2-019-1499-8

[93] Mandal S, Mondal S. Probabilistic approaches and landslide susceptibility. Geoinformatics and modeling of landslide susceptibility and risk. Environmental science and engineering. Springer book series (ESE). 2019:145-163

[94] Mezughi TH, Akhir JM, Rafek AG, Abdullah I. Landslide susceptibility assessment using frequency ratio model applied to an area along the E-W Highway (Gerik-Jeli). Am J Environ Sci. 2011;7:43-50

[95] Oh HJ, Lee S, Wisut C, Kim CH, Kwon JH. Predictive landslide susceptibility mapping using spatial information in the Pechabun Area of Thailand. Environ Geol. 2009;57:641-651

[96] Silalahi FES, Pamela YA, Hidayat F. Landslide susceptibility assessment using frequency ratio model in Bogor, West Java. Indonesia. Geosci. Lett. 2019;6:10

[97] Zine El Abidine, R., Abdel Mansour, N. Landslide susceptibility mapping using information value and frequency ratio for the Arzew sector (Northwestern of Algeria). Bulletin of the Mineral Research and Exploration. 2019;160:197-211. https://doi.org/10.19111/bulletinofmre.502343

[98] Haoyuan H, Chen W, Xu C, Youssef AM, Pradhan B, Bui DT. Rainfall-induced landslide susceptibility assessment at the Chongren area (China) using frequency ratio, certainty factor, and index of entropy. Geocarto International. 2016. DOI: 10.1080/10106049.2015.1130086

Section 7

Landslide Warning

Chapter 12

Detection and Warning of Tsunamis Generated by Marine Landslides

Mal Heron

Abstract

Seismic signals provide an effective early detection of tsunamis that are generated by earthquakes, and for epicentres in the hard-rock subduction zones there is a robust analysis procedure that uses a global network of seismometers. For earthquakes with epicentres in soft layers in the upper subduction zones the processes are slower and the seismic signals have lower frequencies. For these soft-rock earthquakes a given earthquake magnitude can produce a bigger tsunami amplitude than the same earthquake magnitude in a hard rock rupture. Numerical modelling for the propagation from earthquake-generated tsunamis can predict time of arrivals at distant coastal impact zones. A global network of deep-water pressure sensors is used to detect and confirm tsunamis in the open ocean. Submarine landslide and coastal collapse tsunamis, meteo-tsunamis, and other disturbances with no significant seismicity must rely on the deep-water pressure sensors and HF radar for detection and warning. Local observations by HF radar at key impact sites detect and confirm tsunami time and amplitude in the order of 20–60 minutes before impact. HF radar systems that were developed for mapping the dynamics of coastal currents have demonstrated a capability to detect tsunamis within about 80 km of the coast and where the water depth is less than 200 m. These systems have now been optimised for tsunami detection and some installations are operating continuously to provide real-time data into tsunami warning centres. The value of a system to warn of hazards is realised only when coastal communities are informed and aware of the dangers.

Keywords: tsunami, marine landslides, hazards, warnings, HF radar

1. Introduction

The phenomenon of 'tsunami' occurs very often in large bodies of water around the world. The great majority of these are small, even unnoticeable, but have the physical characteristics of a shallow-water gravity wave with periods 10–40 minutes, which define a tsunami. The recording of tsunamis has historically been based on the amount of damage to coastal communities and the magnitude of a submarine earthquake on the moment magnitude scale (M_w) which has been developed from the Richter Scale [1]. These are effectively logarithmic energy scales. Neither of these metrics relate well to the amplitude of the associated tsunami wave in the open ocean, which is a more reliable metric because a medium-scale earthquake ($M_w = 7$) in a landslide earthquake can generate the same tsunami amplitude as a severe ($M_w = 9$) earthquake in a deeper hard-rock subduction zone.

Because of the absence of a standard, the records of 'notable' tsunamis vary among authors, but a general consensus emerges that there is one major tsunami about every 3 years, of which about 75% are caused by earthquakes originating from megathrusts in hard rock. The remaining 25% are mostly landslide tsunamis. The genesis of tsunamis varies with locations and while any large water body like lakes and inland seas are susceptible, it is the so-called ring of fire around the Pacific that has the highest tsunami occurrence of about 80% [2].

Tsunami warning methods fall into three categories. The first arises from the analysis of seismic data collected in the region of the earthquake epicentre. Seismic signals recorded on seismometers located on land near to an earthquake epicentre are used to report estimates of the magnitude and location within a few minutes of the rupture [3]. This can be extended to differentiating between landslide tsunamis and hard-rock megathrust tsunamis [4, 5] by analysing the periodicity in the seismic signals.

The second warning category is obtained through ocean observations in the deep ocean. A network of DART ((Deep-ocean Assessment and Reporting of Tsunamis) moorings consist of a benthic pressure sensor to detect small, but sustained, changes in water depth and a surface buoy for communications [6]. DART moorings can detect tsunamis with amplitudes greater than about 3 cm, and immediately transmit an alert via a satellite link. A network of DART moorings is coupled with seismometers and numerical modelling to warn of potential tsunami impacts around the coastal boundaries of that ocean basin.

The third warning category is at the site of potential impact. Observations of tsunamis approaching in the shallow water on the continental at a critical site give alerts that are accurate in timing and amplitude, but are issued typically less than one hour before impact. The main value of these technologies is to confirm alerts if they have already been given from the epicentre location and the mid-ocean systems, and also to issue warnings for tsunamis generated in the local area. The most promising technology in this category is HF ocean radar that can detect an approaching tsunami at a range of about 100 km, or at the edge of the continental shelf if that is closer than 100 km. The resolution of DART technology and HF radars are consistent with the suggestion that a tsunami wave with an amplitude greater than 0.03 m in the deep ocean should be considered potentially hazardous when it impacts the coast.

Included in this third category of warnings at the site of impact is a cultural awareness of local people to look at the ocean and understand visible changes. For example, any list of 'notable' historic tsunamis recorded would start with a report by Herodotus in 479 BC during the Persian siege of the town of Potidaea (reported by [7]) as "a great flood-tide, higher, as the people of place say, than any one of the many that had been before" which obliterated the Persians who thought they had taken a strategic advantage of the preceding retreat of the water. Herotodus had written the first record of a tsunami impact. Up to half of tsunami impacts on the coast have an initial draw-down which (with suitable education) serves as an excellent warning for local people. Local warnings like HF radar are imperative when the tsunami approaches as a crest.

The well-established global network of seismometers can produce location and magnitude estimates within a few minutes of the event. Based on these data, numerical modelling (e.g. [8]) is used to forecast arrival times of any resulting tsunami at coastal sites around the world using properties of tsunamis which propagate as shallow water gravity waves (even in the deepest oceans!) with a velocity given by

$$c = \sqrt{gh} \tag{1}$$

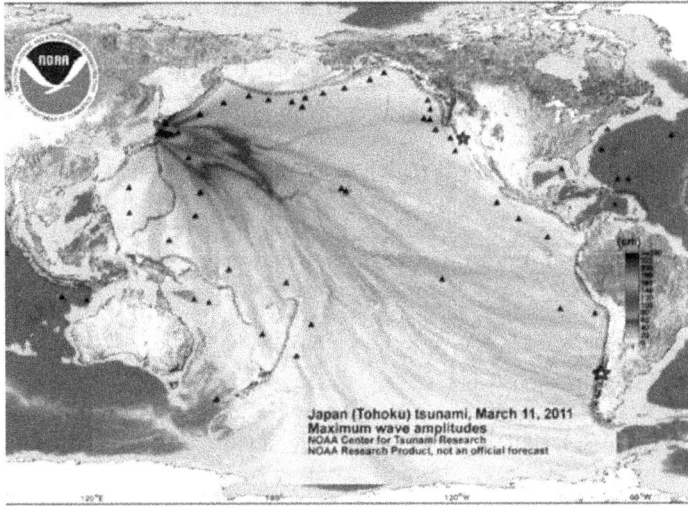

Figure 1.
*Maximum tsunami amplitudes calculated by the MOST model for 24 hours following the magnitude 9.0
earthquake near Tohuko, Japan on 11 march 2011. The faint grey contours show the estimated times of arrival.
http://nctr.pmel.noaa.gov.*

where g is gravitational acceleration and h is the depth of the water column.
At h = 3000 m the tsunami speed is about 170 ms^{-1} and on a continental shelf of
depth 50 m it is about 30 ms^{-1}. This dramatic slowing-down near the coast raises
the value of local observations at impact sites.

The propagation characteristics of the Tohuko Earthquake 2011 shown in
Figure 1 are calculated by the MOST (Method of Splitting Tsunami, [8]) model for
the Pacific Ocean following the earthquake with epicentre 29 km deep and 130 km
from the east coast of Honshu. This computation is a heavy load and, in practice
many warning centres have a library of scalable forecasts for tsunamis that are pre-
calculated for a range of magnitudes and epicentres at regular spacing (say 100 km)
along likely fault zones.

The most critical place for rapid warnings is the adjacent coast which for the
Tohuko earthquake, had tsunami impact approximately 84 minutes after the seismic signals. This is a typical warning time in the local region where the tsunami
amplitudes are greatest (**Figure 1**). Deep water DART Buoys in the Pacific Ocean
can confirm the magnitude of the tsunami, with appropriate delays in the order of
hours (**Figure 1**). Arrival of the tsunami at all impact zones can be confirmed by
local observations and warnings. In most cases the local confirmation of an imminent tsunami would be issued as a follow-up on prior alerts for the event, but for
landslide tsunamis, coastal collapse tsunamis and other non-seismic tsunamis the
local observations may be the only way to give the primary warning.

2. Seismic signal warnings

Seismometers provide the traditional data for the estimation of magnitude and
location of the epicentre of an earthquake, and a global network of instruments
provides rapid and reliable information. The development of seismometry has
traditionally been focused on earthquakes from megathrusts of hard rock in the
subduction zones, but recent work has been reported on ruptures in shallow, soft

rock subduction zones and submarine landslides which produce seismic signals that have different characteristics.

2.1 Seismic signals from tsunamigenic earthquakes

Data from seismometers in the region of an earthquake have been the traditional means of issuing tsunami warnings. So-called Tsunamigenic Earthquakes result mainly from shearing movement at the tectonic plate boundaries and volcanic hot spots in the lithosphere. Tsunamigenic earthquakes typically occur when there are vertical as well as horizontal components in megathrusts on fault lines in the hard rock deep in subduction zones beneath the ocean floor. The energy given to a resulting tsunami comes from the potential energy released during the seismic thrust. The original Richter Scale for earthquake magnitude is illustrated in **Figure 2** from Richter's book [1] where the maximum amplitude of the P waves, and the delay between S and P signals are used to determine the earthquake magnitude.

The relationship between Richter's magnitude and the rupture is given by Aki [9] as:

$$M = \mu AD \tag{2}$$

where D is the slip, μ is the rock rigidity, and A is an area equal to $D \times W$, where W is the depth of the fracture.

Richter's method did not take account of the spectrum of components in the seismic signal, and saturates when $M > 8$. Kanamori [10] considered a range of spectral components in the seismic signal to define a Moment Magnitude, M_w, which agrees with Richter's magnitude for small earthquakes, is accurate for $M_w > 9$, and is now widely used to specify earthquake magnitudes (even though it is often

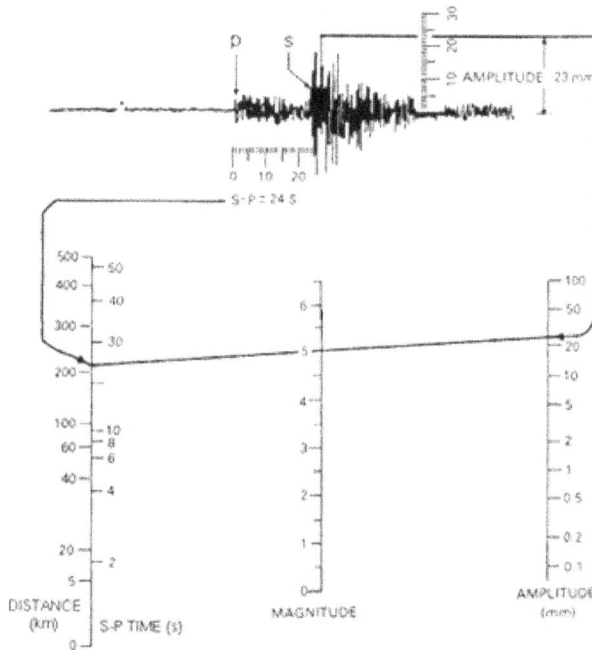

Figure 2.
Richter's relationship between the seismic signals and the logarithmic Richter scale. From Richter, Elementary Seismology [1].

called the Richter Scale). The simplest approach for tsunami warnings is that if M_w > 8.0 and the epicentre is offshore then it is likely that a tsunami will be generated.

Working towards a strategy to provide rapid local tsunami warnings, Melgar et al. [3] use scaling relationships

$$log_{10}D = -2.37 + 0.57M_w \tag{3}$$

$$log_{10}W = -1.86 + 0.46M_w \tag{4}$$

to estimate the width and length of an earthquake deformation based on M_w. Then, using the predefined slab model of Hayes et al. [11] they estimate horizontal and vertical deformations of the sea floor, and the magnitude of resulting tsunamis. This strategy provides estimates of tsunami genesis from hard rock ruptures that are sufficiently accurate to provide tsunami warnings. This method is shown to deliver warnings within a few minutes of the rupture and is the basis of warning systems in Japan, Indonesia and Australia [12, 13]. For the 2011 Tohoku earthquake the rapid estimate gave M_w = 9.3 when the final value was calculated at M_w = 9.0. The subsequent propagation of the tsunami is shown in **Figure 1** which is calculated by the MOST model.

2.2 Seismic signals from landslide earthquakes

The name Tsunami Earthquake was coined by Kanamori [14] and does not include the Tsunamigenic Earthquakes discussed in the previous section. These are tsunamis that are significantly bigger than one would predict from the seismic data, and are generated by deformation in the soft rock in the upper subduction layer or by a submarine landslide in a thick, stratified sedimentary layer on a bathymetric slope.

The Mentawai earthquake off the west coast of Sumatra, Indonesia on 25 October 2010 had a medium magnitude of 7.8 but produced a large tsunami that caused significant coastal damage and loss of over 400 lives. This tsunami was significantly greater than would normally be expected from an earthquake of that magnitude. Analysis of the seismic records [4, 5] showed that the Mentawai earthquake was a result of slow deformation in the upper layers of the subduction zone.

Earlier work by Kanamori [14] had shown that weak earthquakes with slow deformation time constants could produce significant tsunamis. This work was done using data from the Aleutian Islands earthquake of 1946, and the Sanriku earthquake of 1896, both of which were relatively weak earthquakes that produced very large tsunamis. Slow deformation, of around 100 s, does not generate high frequency seismic signals like those shown in **Figure 2** and Kanamori's conclusion is that the abnormally slow deformation at the source of the earthquake generated the tsunami. This is consistent with tsunamis from submarine landslides following ruptures in the weakly coupled soft rock layer on the inner margins of ocean trenches.

Sahakian et al. [15] compared the seismic signals from six earthquakes of similar magnitude of 7.6–7.9, which were chosen because GPS data on earthquake amplitudes were available from GNSS recordings, as well as a local seismometer station. The six earthquakes were Ibaraki, Japan 2011; Nicoya, Costa Rica 2012; Iquique, Chile 2014; Melinka, Chile 2016; and Mentawai, Indonesia 2010. In **Figure 3** the acceleration from the local seismometer, and the vertical displacement time series are shown on the same ordinate scale.

From these data, Sahakian et al. [15] confirmed that the Mentawai earthquake in the soft rock in the upper levels of the subduction zone did not generate the

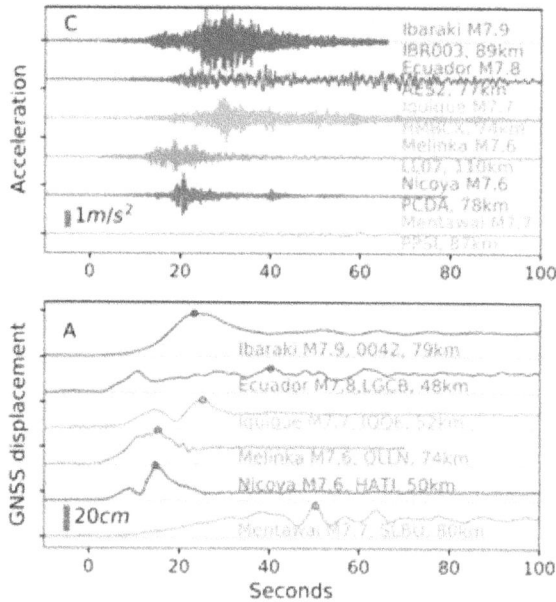

Figure 3.
Data from five earthquakes with similar magnitudes in the 7.6–7.9 range. The earthquake name and magnitude, the seismometer station name and its distance from the epicentre are given for each event. The Mentawi earthquake, and to a lesser extent Ecuador, have smaller accelerations in the seismic signal but comparable vertical displacements in the GNSS data. From Sahakian [15] USGS doi.org/10.3133/circ1187.

high-frequency seismic signals that are generated by megathrusts of earthquakes in the hard rock deeper in the zone. If the slower fluctuations are accompanied by large amplitudes, then it is concluded that the amplitude-to-energy ratio can be used to detect a tsunami earthquake when the magnitude is lower. By comparing fluctuation amplitudes (observed by GPS) with earthquake energy, Sahakian showed that it is possible to issue an alert for a potential tsunami from a tsunamigenic earthquake.

Sahakian et al. [15] suggested a method for early warnings of Tsunami Earthquakes is to estimate M_{PGA} from the seismometer and M_{PGD} from the GNSS vertical displacements. Then a low M_{PGA} coupled with a high M_{PGD} suggests that the event has ruptured soft and compliant rock high in the subduction zone with a high likelihood of producing a large tsunami.

3. Open Ocean observations

Observations in the open and deep ocean are used to give tsunami warnings to locations in the ocean basin that are a long way from the earthquake epicentre. These warnings are relevant to the most severe earthquakes because of the attenuation and geometric spreading of tsunami waves across an ocean basin as shown in **Figure 1** for the 2011 Tohuku earthquake and tsunami.

Even the largest of destructive tsunamis have relatively small amplitudes of up to a few tens of centimetres in deep water. This is illustrated by the altimeter data recorded by the JASON-1 satellite in an opportunistic transit 2 hours after the Sumatra-Andaman earthquake 26 December 2004 [16]. The altimeter recorded a maximum water elevation of about 50 cm in open ocean compared with reports of elevations up to 30 m at some coastal impact points. A DART buoy in the Bay of

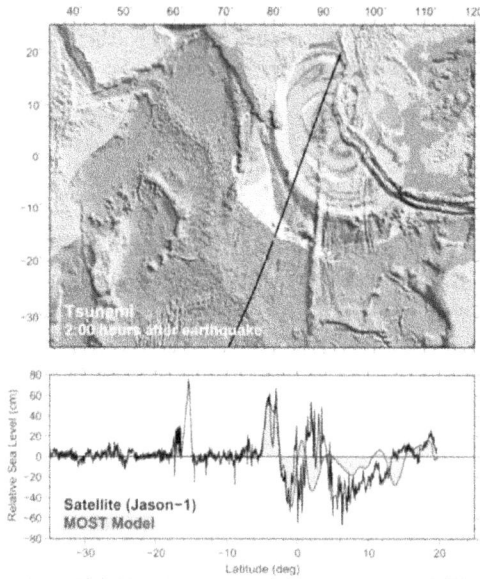

Figure 4.
Transit path of Jason-1 altimeter superposed on estimated elevations from the MOST model 2 hours after the Sumatra-Andaman 2004 earthquake. The maximum water elevation was about 50 cm. From Gower [16] Taylor and Francis tandfonline.com.

Bengal would have given operational confirmation of the timing and warning of the scale of the tsunami.

DART buoys consist of a bottom-mounted pressure sensor with a cable connection to a surface buoy which communicates to a monitoring laboratory through the Iridium network. The pressure sensor takes 15-sec time series at sensitivity of 1 mm of sea water and filters out the high frequencies. If two successive 15-sec averages exceed a projection from the past 3 hours by more than 3 cm the system goes into rapid reporting mode to send data every minute [17, 18]. Following the Sumatra-Andaman earthquake NOAA/PMEL developed an ETD (Easy to Deploy) upgrade to the DART system [19] and this technology is being adopted widely (**Figure 4**) to enable national tsunami warning centres to improve warning systems. The DART system is robust and makes a significant contribution to tsunami warning. DART buoys provide a critical element of the global tsunami warning capability but need to be complemented by seismic and GPS systems for regional warnings near to the epicentre, and systems for warnings in local impact areas whether they are near to the epicentre or distant across oceans.

4. Local impact observations

Apart from visual observations at the beach, the only real-time technology for imminent impacts of tsunamis is land-based HF ocean radar. Observations can only be made over shallow waters <200 m deep in coastal waters, which can give warnings typically 20–60 minutes before impact. To be effective as a warning method HF radar needs to be supported as much as possible with seismic-based alerts or warnings from DART buoys. For tsunamis generated in the local region, especially landslide or coastal collapse events, the local warning from an HF radar may be the only alert possible.

A significant feature not illustrated in **Figure 1** is the growth in amplitude of a tsunami as it slows down in shallow water. As a first-order approximation this growth can be estimated following Green [20] as

$$a(d) = a(D)(D/d)^{\frac{1}{4}} \tag{5}$$

where a is tsunami amplitude, d is water depth, and D is a reference (deep water) depth. An example of this phenomenon is illustrated by **Figure 5** which shows the amplitude of the Sumatra-Andaman tsunami as about 50 cm in the open water of Bengal Bay, when the tsunami later rose to near 30 m in some impact areas. Associated with the amplification is an enhancement of the velocities of water particles in the propagating wave. The circulating water particles in a gravity wave are manifest on the surface as the to-and-fro motion that can be observed as a swell wave propagates past a point on the ocean. The increase in the maximum to-and-fro velocity, v_m, is [21];

$$v_m(d) = v_m(D)(D/d)^{\frac{3}{4}}. \tag{6}$$

The primary product of land-based HF ocean radars is surface currents mapped at high spatial resolution over the coastal ocean, and over 400 systems have been installed around the world for that purpose [22]. The potential for HF radars to observe tsunamis was suggested by Barrick [23] and confirmed when several HF radars in Japan as well as North and South America recorded signals as the tsunami from the Tohuko 2011 earthquake reached the west coast of the Americas [24, 25]. In these cases the radars were configured for currents in coastal circulation dynamics, and following the events of March 2011 there was a focus on optimising HF radars for real-time tsunami observations by measuring $v_m(d)$ in Eq. (6).

There are two main HF radar technologies that are widely available for mapping surface currents in coastal waters. Both radar systems operate by receiving radar echoes from the rough, conducting sea surface, and both technologies use timing to define the range of a target zone on the ocean. But they have quite different solutions for determining the angle in the (r,θ) plane. One is the Seasonde system which uses wide angle crossed-loop receiving antennas to define the pointing direction of the radar [23]; and the other is the WERA system which uses a phased

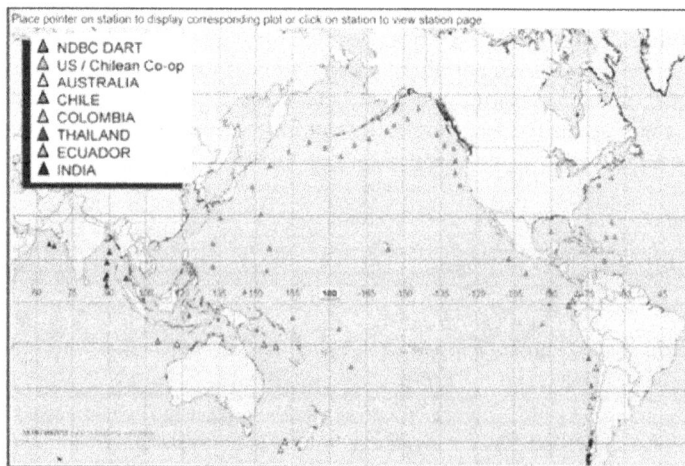

Figure 5.
Global network of DART buoys. www.ndbc.noaa.gov/dart.shtml accessed 21 July 2021.

array of elements as the receive antenna to define azimuth [26]. This differentiation leads to quite different solutions to the challenge of issuing warnings for imminent tsunamis.

The Seasonde software offers a q-factor calculation developed empirically from past tsunami observations and simulations [27, 28]. The observation area is partitioned into strips 2 km wide running parallel to the benthic contours. When the radial current components in three adjacent strips are highly correlated, are showing a trend in magnitude, and are significantly different from the background current, the q-factor index is incremented. **Figure 6** shows the q-factor calculated for data from a Seasonde radar at Point Estero, California (marked with a solid 'x' in **Figure 1**) following the Tohuko 2011 earthquake. Time series of the average current in each strip are taken every 4 minutes, and **Figure 6** shows the data over three strips with the q-factor calculated from the three strips in the 8–14 km range. Tsunami warnings are issued when the q-factor exceeds a trigger level that is set for the conditions prevailing at the specific site, but typically the trigger level is q = 500. Note that the data shown in **Figure 6** were taken from a radar installation optimised for current dynamics, and not for tsunamis. It is a proof of concept.

A WERA station at Rumena in Chile (marked with an 'x' in **Figure 1**) also recorded the tsunami from the Tohuko, 2011 earthquake. In **Figure 7** the colours show current anomaly, with the background removed, with time and range for the beam in the NW direction. From these data and simulations, Gurgel et al. [30] and Dzvonkovskaya et al. [29] developed a probability approach where a time series is

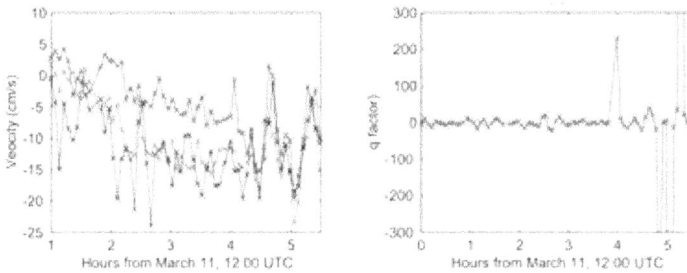

Figure 6.
SeaSonde time series of onshore velocity components and q-factors at point Estero (a) blue: 8–10 km offshore; red: 10–12 km offshore, and black: 12–14 km offshore; (b) q-factor for the 8–14 km interval. From Lipa et al. [25].

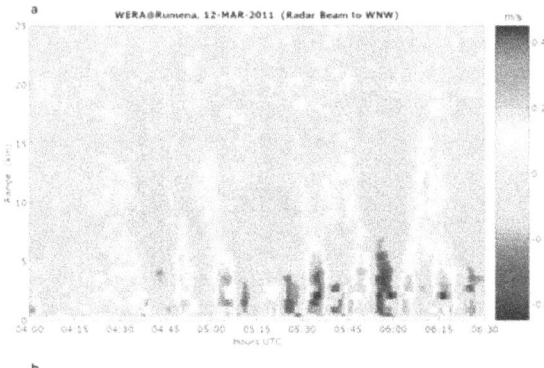

Figure 7.
Surface currents on a time vs. range visualisation for the Tohuko 2011 tsunami approaching the coast at Rumena, Chile. From Dzvonkovskaya et al. [29].

taken over 133 s for each grid point over the mapped area and a probability of an anomalous current (compared with background currents) is assessed at each point. A probability map is produced from 133-second overlapping time series and issued every 33 s. An example of a probability map is shown in **Figure 8** for the data set from the Tohuko 2011 tsunami taken at Rumena at 0545 UT, some 45 minutes before impact. Note that the data shown in **Figures** 7 and **8** were taken from a radar installation optimised for current dynamics, and not for tsunamis. It is a proof of concept.

Dzvonkovskaya et al. [29, 31] further developed this method into a robust algorithm for an estimation of the 'Probability of Tsunami' for the site. The final tsunami warning product is produced by statistical processing of successive 2D probability maps. An example of the alerts and warnings issued in real time for a WERA system configured for tsunami warnings is shown in **Figure 9**. This event

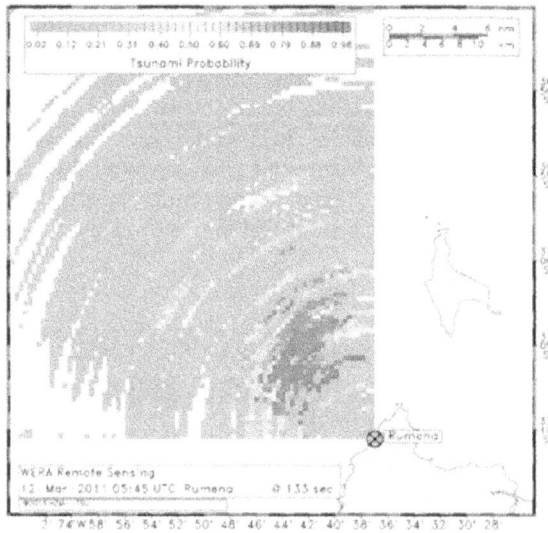

Figure 8.
Map of estimates of tsunami probability for independent grid points from the WERA radar at Rumena following the Tohuko 2011 earthquake. From Dzvonkovskaya et al. [29].

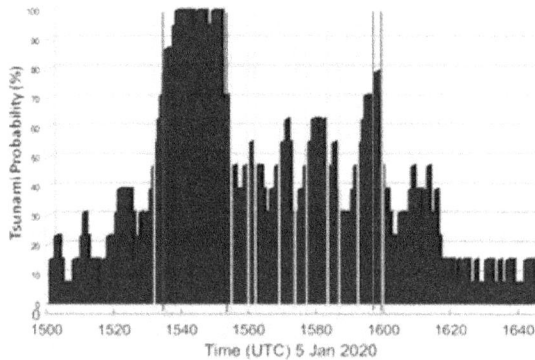

Figure 9.
Tsunami probability for the WERA radar for Tofino, Canada. Surface currents from the whole grid are combined to give a single probability index that is issued every 33 s in real time. 'ATTENTION' is issued when the tsunami probability exceeds 50% (between yellow lines) and 'ALERT' is issued at 75% (between red lines) in real-time.

Figure 10.
WERA HF radar stations at Tanjung and Kahai deployed for detecting tsunamis in the irregular pentagon inside the yellow lines. This is a part of the GITEWS for warning of tsunamis generated in the Krakatoa caldera. Base image GoogleEarth.

was a tsunami-like disturbance produced by a severe meteorological front at Tofino, British Columbia, Canada in 2020. The Tsunami Probability (TP) bulletins are issued every 33 seconds to the host system. The recommended warning levels are issued as 'attention' if $50 < TP < 75$, and as 'alert' if $TP > 75$ as shown in **Figure 9** for the event at Tofino.

HF radars have been deployed at several places with the primary purpose of tsunami warning where data are returned in real time to a tsunami warning centre. One of these is in the Sunda Strait where the volcano island Anak Krakatau appeared above the sea in 1927 on the edge of the Krakatoa Caldera formed in 1883. After several days of seismic activity in December 2018, it erupted with an area of about 64 hectares and volume of 0.2 km³ collapsing into the sea generating a tsunami that resulted in 437 fatalities and over 30,000 injuries on the adjacent Java and Sumatra islands [32]. The collapse had no strong short-period seismic signals and did not provide a seismic tsunami warning. In view of the volcanic activity in the area, the German Indonesian Tsunami Early Warning System (GITEWS) was established as an integrated system for warning of locally generated tsunamis with sensors including seismic, acoustic and HF radar [33]. The configuration of the HF radar stations is shown in **Figure 10**. This radar installation is being used primarily in the GITEWS system but is also producing maps of surface currents over the area outlined by the irregular pentagon in **Figure 10**. There is a strong interest in the ocean dynamics in the Sunda Strait because the long-term flow-through of warm water from the Java Sea to the Indian Ocean is a key driver of ocean circulation. This WERA system produces current maps on a 1×1 km grid every 20 minutes for circulation applications, and evaluate tsunami activity every 33 s on a continuous schedule.

5. Conclusion

Tsunamis generated by hard-rock megathrust earthquakes, like Tohuko 2011 in the Pacific and Sumatra-Andaman 2004 in the Indian Ocean, give high-frequency seismic signals from which reliable, fast estimates of tsunami amplitudes are produced, and propagation modelling across neighbouring oceans can be made, with reliable confirmation from deep-ocean DART buoys. For these tsunamis, confirmation at key coastal sites is useful because shallow-water bathymetry and coastal topography are significant parameters for the terrestrial run-up of water. The warning systems for tsunamis generated by hard-rock megathrust earthquakes are robust and are widely used. Routine monitoring is being used at coastal sites that have infrastructure or populations at risk. The use of HF radar at key at-risk sites gives final confirmation of amplitude and timing before tsunami impact.

Tsunamigenic soft-rock earthquakes in the upper subduction zone have slow-response seismic signals that can lead to underestimation of resulting tsunamis if hard-rock algorithms are used. There is active development of methodology to estimate the amplitudes of tsunamis from tsunamigenic earthquakes using regional seismic and GPS signals. Data from the Mentawai 2010 earthquake and tsunami have provided a foundation for the development of this method. Implementation requires installation of GPS (GNSS) monitoring of earthquake amplitudes in regions where there are known unstable sedimentary bedforms. Tsunamis from tsunamigenic earthquakes are detected on DART buoys and local monitoring at key at-risk sites on the coast gives confirmation. Local monitoring may be the primary warning at impact sites close to the epicentre.

Submarine landslides and coastal collapse have produced damaging tsunamis with impacts mostly confined to the local region. These tsunamis have no seismic warnings. Other tsunami genesis mechanisms without seismicity include glacial calving, which is localised, and meteotsunamis which are generated by meteorological fronts and similar in scale to storm surges at the coast. The seismic and DART technologies are less applicable to these events because most damage has been in the source region and local monitoring takes on more urgency. HF radar is a proven technology for tsunami detection in shallow coastal waters where warnings can be issued in the order of 20–60 minutes before impact (depending on the width of the shallow coastal shelf). While HF radars have been installed primarily for tsunami detection and warning in several regions, they can also provide maps of coastal currents for research and management of coastal industries like ports, marine reserves and coastal engineering. There are over 400 HF radar installations worldwide [22], usually in regions of high economic or social interest. A useful strategy in the short term would be to retro-fit existing HF radars that have tsunami detection capability, with tsunami detection software that is integrated into central tsunami warning hubs.

Acknowledgements

I am grateful to R. Gomez and to reviewers for comments that improved the manuscript. The author has no known conflicts. Commercial radars Seasonde and WERA are mentioned: the reader is referred to the relevant web sites for full technical specifications.

Detection and Warning of Tsunamis Generated by Marine Landslides
DOI: http://dx.doi.org/10.5772/intechopen.99914

Author details

Mal Heron
Marine Geophysics Laboratory and Physical Sciences, James Cook University,
Australia

*Address all correspondence to: mal.heron@ieee.org

IntechOpen

References

[1] Richter, C.F., Elementary Seismology, 768 pp., 205 illus. W. H. Freeman and Company, San Francisco, and Bailey Bros. and Swinfen Ltd., 75 London, 1958. www.US.macmillan.com

[2] National Geographic Society. Tsunamis. National Geographic. Accessed March 1, 2014. http://environment.nationalgeographic.com/environment/natural-disasters/tsunami-profile/.

[3] Melgar, D., Allen, R. M., Riquelme, S., Geng, J., Bravo, F., Baez, J. C., et al. (2016). Local tsunami warnings: Perspectives from recent large events. Geophysical Research Letters, 43, 1109–1117. DOI:10.1002/2015GL067100

[4] Lay, T., C.J. Ammon, H. Kanamori, Y. Yamazaki, K.F. Cheung and A.R. Hutko, The 25 October 2020 Mentawai tsunami earthquake (Mw 7.8) and tsunami hazard presented by shallow megathrust ruptures, Geophys. Res. Letters, 38, 5pp, L06302, 2011. DOI: 10.1029/2010GL046552.

[5] Newman, A.V., G. Hayes, Y. Wei and J. Convers, The 25 October 2010 Mentawai tsunami earthquake, from real-time discriminants, finite-fault rupture, and tsunami excitation, Geophys. Res. Letters, 38, L05302, 2011. DOI:10.1029/2010GL046498.

[6] Bernard, E., and C. Meinig (2011): History and future of deep-ocean tsunami measurements. In Proceedings of Oceans' 11 MTS/IEEE, Kona, IEEE, Piscataway, NJ, 19–22 September 2011, 15 No. 6106894, 7 pp.

[7] Smid, T. C. 'Tsunamis' in Greek Literature. Greece and Rome, 2nd Ser., Vol. 17, No. 1 (April 1970), pp. 100–04 (102f.)

[8] Titov, V. V., F. I. Gonzalez, E. N. Bernard, M. C. Eble and H. O. Mofjeld, Real-time tsunami forecasting: Challenges and solutions, in Developing Tsunami-resilient Communities, ed E. N. Bernard, pp. 41-58, Springer, Netherlands, 2005

[9] Aki, K., Earthquake Mechanism, Tectonophysics, 13, 423-446, 1972 do: 10.1016/0040-1951(72)90032-7

[10] Kanamori, H. The energy release in great earthquakes, J. Geophys. Res., 82 (20): 2981–2987, 1977. doi:10.1029/jb082i020p02981.

[11] Hayes, G. P., D. L. Wald, and R. L. Johnson (2012), Slab1.0: A three-dimensional model of global subduction zone geometries, J. Geophys. Res., 117, B01302, DOI:10.1029/2011JB008524.

[12] Allen, S. C. R., and D. J. M. Greenslade (2008), Developing tsunami warnings from numerical model output, Nat. Hazards, 46(1), 35–52. 06

[13] Hoshiba, M., and T. Ozaki, Earthquake early warning and tsunami warning of the Japan Meteorological Agency, and their performance in the 2011 off the Pacific Coast of Tohoku Earthquake (M_w 9.0), in Early Warning for Geological Disasters, eds F. Wenzel and J. Zschau, pp. 1-28, Springer, Berlin, 2014.

[14] Kanamori, H. (1972). Mechanism of 82 tsunami earthquakes. Physics of the 83 Earth and Planetary Interiors. 6 (5): 84 346–359. DOI:10.1016/0031-9201 (72) 85 90058-1

[15] Sahakian, V. J., Melgar, D., and Muzli, M., Weak near-field behavior of a tsunami earthquake: Toward real-time identification for local warning. Geophys. Res. Letters, 46, 2019. DOI: 86 10.1029/2019GL083989

[16] Gower, J. (2007). The 26 December 2004 tsunami measured by satellite

altimetry, Int. J. Remote Sens., 28, 2897–2913, DOI:10.1080/01431160601094484. Taylor and Francis www.tandfonline.com.

[17] Meinig, C., S.E. Stalin, A.I. Nakamura, F. González, and H.G. Milburn (2005): Technology Developments in Real-Time Tsunami Measuring, Monitoring and Forecasting. In Oceans 2005 MTS/IEEE, 19–23 September 2005, Washington.

[18] Milburn, H.B., A.I. Nakamura and F. I. Gonzalez, Real-Time Tsunami Reporting from the Deep Ocean, Proc. MTS/IEEE OCEANS, September 1991.

[19] Lawson, R.A., D. Graham, S. Stalin, C. Meinig, D. Tagawa, N. Lawrence-Slavas, R. Hibbins, and B. Ingham (2011): From Research to Commercial Operations: The Next Generation Easy-to-Deploy (ETD) Tsunami Assessment Buoy. In Proceedings of Oceans'11 MTS/IEEE, Kona, IEEE, Piscataway, NJ, 19–22 September 2011, No. 6107114, 8 pp

[20] Green G, On the Motion of Waves in a Canal of Variable Depth G Green - Cam. Phil. Trans., VI: 1837. 457 p.

[21] Kinsman, B., Wind Waves, Prentice-Hall, Englewood Cliffs, NJ, USA, 1965.

[22] Roarty H, Cook T, Hazard L, George D, Harlan J, Cosoli S, et al., 2019. The Global High Frequency Radar Network. Front. Mar. Sci. 6:164. DOI: 10.3389/fmars.2019.00164

[23] Barrick, D., A coastal radar system for tsunami warning, Rem. Sens. Env., 8, 353-358, 1979 DOI:10.1016/0034-4257 (79)90034-8

[24] Dzvonkovskaya, A., Ocean surface current measurements using HF radar during the 2011 Japan tsunami hitting Chilean coast. In: Proc. of IEEE IGARSS20 2012, Munich, Germany,

2012, pp. 7605-7608, DOI:10.1109/IGARSS.2012.6351867.

[25] Lipa, B., J. Isaacson, B. Nyden and D. Barrick, Tsunami arrival detection with high frequency (HF) radar, Remote Sens. 2012, 4(5), 1448-1461; DOI:10.3390/rs4051448

[26] Gurgel KW, Antonischki G, Essen HH, et al. Wellen radar (WERA): A new ground-wave HF radar for ocean remote sensing. Coastal Engineering. 1999;37: 219–234.

[27] Lipa, B., D. Barrick and J. Isaacson, Coastal tsunami warning with deployed HF radar systems, Chapter 5 in Tsunami, Mohammad Moktari (Ed.), InTech. 2016. DOI:10.5772/63960. Available Online: http://www.intechopen.com/books/tsunami/coastal-tsunami-warning-with-deployed-HF-radar-systems

[28] Lipa, B., H. Parikh, D. Barrick, H. Roarty, and S. Glenn. High frequency radar observations of the June 2013 US East Coast Meteotsunami, Nat Hazards (2014) 74:109–122 DOI 10.1007/s11069-013-0992-4.

[29] Dzvonkovskaya, A., L. Petersen, T. Helzel, and M. Kniephoff, High Frequency Ocean radar support for tsunami early warning systems, Geosci. Res. Letters, 5:29, 2018. DOI:10.1186/s40562-018-0128-5.

[30] Gurgel K-W, Dzvonkovskaya A, Pohlmann T, Schlick T, Gill E (2011) Simulation and detection of tsunami signatures in ocean surface currents measured by HF radar. Ocean Dynamics, Springer. DOI:10.1007/s1023 6-011-0420-9

[31] Dzvonkovskaya, A., 2018. HF surface wave radar for tsunami alerting: From system concept and simulations to integration into early warning systems, IEEE A&ES Mag 33:48–58. DOI: 34 10.1109/MAES.2018.160267

[32] Ye, L., H. Kanamori, L. Rivera, T. Lay, Y. Zhou, D. Sianipar and K. Satake, The 22 December 2018 Tsunami from Flank Collapse of Anak Krakatau Volcano during Eruption, Science Advances, 6/3, eaaz1377, 2020. DOI: 10.1126/sciadv.aaz1377

[33] Gomez, R., T. H. Tran, A. Ramdhani, and R. Triyono, HF Radar Validation and Accuracy Analysis using Baseline Comparison Approach in the Sunda Strait. Global Oceans 2020: Singapore – U.S. Gulf Coast, 2020, pp. 1-5, DOI: 10.1109/IEEECONF38699.2020.9389158

Empirical Rainfall Thresholds for Landslide Occurrence in Serra do Mar, Angra dos Reis, Brazil

Daniel Germain, Sébastien Roy
and Antonio Jose Teixera Guerra

Abstract

In the tropical environment such as Brazil, the frequency of rainfall-induced landslides is particularly high because of the rugged terrain, heavy rainfall, increasing urbanization, and the orographic effect of mountain ranges. Since such landslides repeatedly interfere with human activities and infrastructures, improved knowledge related to spatial and temporal prediction of the phenomenon is of interest for risk management. This study is an analysis of empirical rainfall thresholds, which aims to establish local and regional scale correlations between rainfall and the triggering of landslides in Angra dos Reis in the State of Rio de Janeiro. A statistical analysis combining quantile regression and binary logistic regression was performed on 1640 and 526 landslides triggered by daily rainfall over a 6-year period in the municipality and the urban center of Angra dos Reis, in order to establish probabilistic rainfall duration thresholds and assess the role of antecedent rainfall. The results show that the frequency of landslides is highly correlated with rainfall events, and surprisingly the thresholds in dry season are lower than those in wet season. The aspect of the slopes also seems to play an important role as demonstrated by the different thresholds between the southern and northern regions. Finally, the results presented in this study provide new insight into the spatial and temporal dynamics of landslides and rainfall conditions leading to their activation in this tropical and mountainous environment.

Keywords: Landslide, rainfall thresholds, quantile regression, tropical environment, Brazil

1. Introduction

Because of their ability to move rapidly, mobilize large amounts of debris, and initiate spontaneously, landslides pose a threat to people and infrastructure [1, 2]. The danger is heightened by the fact that they can occur in wet or dry regions, and on steep or shallow slopes [3]. Synergistic effects between increasing urbanization, sustained deforestation, and increased rainfall variability caused by climate change portend an increase in the frequency of catastrophic landslides such as that experienced in recent years [4].

In Brazil, as elsewhere, the mechanism of slope saturation by rainwater represents the main cause of landslide triggering [5–7]. Their onset is related to exceptional rainfall events of short duration, such as intense precipitation associated with a thunderstorm, or events of long duration and low intensity [8]. Specifically, high-intensity, short-duration precipitation events are known to trigger shallow landslides [9] - landslides with a failure surface depth of less than two meters [10] - while long duration, low intensity precipitation events generally result in deep seated landslides [11]. However, the causal relationship between rainfall and landslides is not so simple [12]. Rather, their initiation is associated with the infiltration of water into the soil that causes an increase in hydraulic pressure and a decrease in resistance, ultimately leading to failure of the affected surface [13]. The effectiveness of the process therefore depends on the hydraulic, physical and mechanical properties of the terrain, in addition to other factors such as slope steepness, vegetation cover and climatic characteristics [12]. In this regard, mountainous regions are particularly favorable to landslides because of the steep slopes, but especially because of the orographic uplift of humid air masses that cause significant precipitation at high elevations and on slopes exposed to prevailing winds [14, 15].

In the Brazilian municipality of Angra dos Reis, located in the Serra do Mar Mountain Range, heavy summer precipitation has historically triggered several landslides, causing a significant number of casualties and considerable damage, especially in the last decade. As an example, the catastrophic events of December 2002 and January 2010 resulted in 93 casualties, forced the evacuation of more than 2,500 residences and generated economic losses of approximately R$120 million. In this tropical environment, the frequency of rainfall-induced landslides is particularly high due to the rugged terrain, heavy summer rainfall, and improper land use regarding the physical and climatic environment [7]. The risk posed by this geomorphological hazard is particularly high due to the fact that almost 60% of the population lives in slope areas [16] and more than 25% (44,000 inhabitants) live where a high risk of landslides is considered [17].

However, since the beginning of the research on rain-triggered landslides in Brazil, no suitable threshold has been proposed for the Angra dos Reis territory. In fact, the thresholds of Guidicini and Iwasa [18] were established for the whole Serra do Mar, whose area is four times larger than that of Angra dos Reis, while the one elaborated by Soares [19] is not representative of the triggering conditions due to the use of an inadequate database. Thus, despite the recurrence of this natural hazard and the socioeconomic risks it poses, the relationships between rainfall characteristics and landslide occurrence remain poorly studied and partly misunderstood in this rugged region of Southeast Brazil. Therefore, the establishment of rainfall thresholds at local and regional scales is of great interest for the municipality of Angra dos Reis and could be an effective tool of prevention and mitigation in the landslide risk management perspective.

2. Regional setting

2.1 Geological and geomorphological settings

The Brazilian municipality of Angra dos Reis is located in the western part of the state of Rio de Janeiro. It represents an area of 825 km^2 composed of four districts (Angra dos Reis, Cunhambebe, Ilha Grande and Mambucaba) comprising 116 neighborhoods. The urban center of the Angra dos Reis district covers an area of 6.5 km^2, or 0.84% of the territory covered by the Angra dos Reis region, and includes 21 neighborhoods.

The geology is composed of 30% granites, 27% orthogneiss, 33% paragneiss with a minor proportion of contemporary sediments (Neogene-Quaternary). In terms of soils in the area, they are characterized by the presence of rock outcrops, fluvio-marine deposits, colluvium, and saprolites [20]. Superficial saprolites are less than two meters thick, have a large amount of boulders, and are generally located on very steep slopes where bedrock outcrops. The thick saprolites, associated with the upper and lower slopes, are more than two meters thick and result from a significant chemical alteration of the rocks in situ favored by the high heat and humidity.

The western part of the state of Rio de Janeiro belongs to the physiographic region of the Serra do Mar, which extends for just over 1,000,000 km along the southern and southeastern Brazilian coastline [21]. The Serra do Mar originated from tectonic movements that began ~80 million years ago (Late Cretaceous) with epirogenic uplift of the crystalline shield throughout southeastern Brazil [22]. Today, this mountain range forms an enormous tectonic barrier parallel to the coast.

Angra dos Reis is located more precisely in the southern part of the Atlantic Plateau, which corresponds to the Bocaína Plateau region and includes the escarpments of the Serra do Mar and the narrow coastal plain of Ilha Grande Bay. In the Bocaína Plateau, the slopes are moderately inclined (10 to 35°), but can exceed 35° in places. The strong geomorphological activity is visible by landslide scars and scree slopes [23]. The urban center of Angra dos Reis, with a summit at 571 m above sea level and very steep slopes, is part of this geomorphological unit. The area of Ilha Grande Bay with its mountainous massif oriented East–West, which culminates at 1031 m of altitude, is also included in this geomorphological unit. The massif has steep slopes, rocky walls, and well incised river channels [23]. Finally, due to the East–West alignment of the Serra do Mar in this area, the slopes are mainly oriented to the North and South, both regionally and locally.

2.2 Climate and land use

The climate is particularly variable due to the proximity of the Atlantic Ocean and the rugged terrain associated with the Serra do Mar. According to the Köppen-Geiger classification, the region is characterized by an Af-type climate; a humid tropical climate without a well-defined dry season corresponding to average monthly temperatures above 18°C and average monthly rainfall above 60 mm. Annual rainfall varies between 2000 and 2500 mm [24]. Typical of tropical regions, there is heavy rainfall in the summer (December to March; with average rainfall exceeding 250 mm and about 16 rainy days/month) and a period of lesser rainfall in the winter (June to August; with total rainfall around 80 mm and less than 10 rainy days/month) [15, 18]. During the rainy season, which concentrates nearly 60% of the annual precipitation [15], rainfall of 200 to 300 mm in 24 to 48 hours is frequent [5]. This intense rainfall is furthermore largely responsible for the high frequency of landslides in this part of Brazil [7, 25].

The intensity and distribution of precipitation is influenced by various static and dynamic factors [26]. Dynamic factors refer to the different air masses and their circulation patterns such as, among others, frontal systems, the South Atlantic Subtropical Anticyclone and the South Atlantic Convergence Zone. The static factors correspond rather to the geographical location (latitude, maritime proximity that facilitates solar radiation, evaporation and cloud formation) as well as to the topographical characteristics (elevation and perpendicular orientation of the Serra do Mar Mountain Range with respect to atmospheric currents that favor the development of intense thunderstorms through the orographic lifting of polar humid air masses blowing in the northwest direction) [24]. Therefore, coastal areas and windward-facing slopes (south-facing) tend to be wetter (2000-2500 mm/ year) due to orographic precipitation, while leeward-facing slopes (north-facing) are generally drier (1400-1700 mm/year) due to moisture loss from advection of air masses over the Mountain Range [27].

The high population growth since the early 1970s in Angra dos Reis, related to the construction of the Governor Mario Covas highway (Br-101) and the Angra 1 and Angra 2 nuclear plants [16], has generated significant pressure on the physical environment. However, the rugged topography (7% plains and 93% hills/mountains), which limits the amount of land available and suitable for human settlement, as well as the lack of land-use planning regulations have caused chaotic development of the territory [28]. This development has led to deforestation, surface sealing, transformation of plateaus into pastures and residential development on steep slopes, generating furthermore an accumulation of waste, a change in natural drainage conditions and anthropic filling and excavation activities that have affected the stability of the slopes and increased the likelihood of landslide occurrence [5]. As a result, the biophysical cover of the municipality of Angra dos Reis, which was once entirely Atlantic Forest (*Mata Atlântica*) in its original state [29], is now much more diverse. Indeed, it is now composed of 86% of secondary Atlantic Forest, with the rest being pastures, urban areas, dunes, mangroves, etc.

3. Data acquisition and methodology

3.1 Landslide database

The landslide inventory includes all the landslides that occurred in the territory of the municipality of Angra dos Reis. The information's included are the geographical coordinates of the landslides and the date of occurrence. However, the inventory does not allow distinguishing between different mass movements (landslide, debris flow, etc.). Therefore, all types of landslides are considered here, without any particular distinction. This is a well-established approach [30, 31] and advantageous considering the fact that the typology of landslides is unknown, unspecified or uncertain since many reports come from citizens, journalists or technicians without adequate scientific training. Duplicate landslides, those with identical geographic coordinates and date of occurrence, were removed from the database. The same is true for cases with erroneous locations. All the recorded landslides were georeferenced and compiled into a geographic database using ArcGIS software [32].

Finally, each of the landslides was associated with a rainfall region according to its location (see Section 3.2), in order to associate or not the cases of landslides with the occurrence of rainfall episodes in the municipality. Subsequently, the landslides were matched to the rainfall data series according to the date of occurrence. This allowed each landslide to be assigned a daily precipitation value (R), 3-, 5-, 10-, 15-, and 30-day antecedent precipitation values, as well as duration (D) and cumulative precipitation values during the rainfall event (E).

3.2 Rainfall analysis

The rainfall data were collected from a regional network consisting of two rain gauges administered by the State Institute of the Environment (SIE) and 19 rain gauges managed by the Civil Defense of Angra dos Reis. In the case of the SIE rain gauges, automatic recordings were made every 15 minutes, 96 times a day. In the case of the rain gauges of the Civil Defense, the data were daily and the reading was done manually every morning at 9:00 am. However, the period covered by the data sets varies considerably depending on the rain gauge station. In order to estimate the amount of rainfall responsible for the occurrence of each of the recorded landslides, the regional study area was first partitioned and an area of influence was calculated for each rainfall station using the Thiessen polygon technique [33] with ArcGIS software [32]. Next, the databases associated with the 21 rainfall stations were agglomerated based on geographic proximity to obtain complete time series for the 6-years period considered.

First, the daily data (R) from each station were associated (or not) with a rainfall event, i.e. a more or less continuous period of rainfall. A rainfall event begins when at least two millimeters of rain have been accumulated in 24 hours and ends at the beginning of a period of at least 24 hours without rainfall. Once the rainfall events were identified, the duration (D) in hours of each episode and the associated total rainfall (E) in millimeters could be calculated. These values were then used to establish thresholds based on the duration of rainfall events (ED).

In a second step, each daily precipitation (R) was associated with antecedent precipitation values. The antecedent precipitation values correspond to the daily totals accumulated over 3, 5, 10, 15 and 30 days before the daily precipitation considered (A (3d), (5d), (10d), (15d), (30d)). These data will allow to evaluate the role of the previous precipitation in the landslide triggering and to determine the most significant previous period.

3.3 Probabilistic rainfall event: duration thresholds (ED) for landslides

The thresholds (ED) were developed from the combination of the variables D and E obtained for each landslide that was triggered during a rainfall event and are defined respectively as the duration (h) and cumulative precipitation (mm) from the start of the rainfall event to the occurrence of the landslide. ED thresholds were developed at the regional and local scales, as well as for the North and South aspects and also for the wet and dry seasons. In the latter case, the considered duration of the dry and wet seasons has been extended to simplify the analyses. Therefore, the dry season is from May to October and the wet season is from November to April.

In all cases, the E (cumulated event rainfall) and D (duration of the rainfall event) values were first plotted in a line graph (log–log coordinates), based on the frequentist method assuming that the threshold curve is a power law such as reported by Guzzetti et al. [34] and Peruccacci et al. [35]:

$$E = \alpha \, D^{\gamma} \qquad (1)$$

where α and γ are the scaling and the shape parameters that control the slope of the power law threshold curve.

The intercept α and the slope γ were then determined through a frequency analysis of the empirical rainfall conditions that have triggered landslides. The large number of landsides recorded over a 6-years period in the study area appears sufficiently complete and representative to determine the 1% and the 5% exceedance probability levels. The mean values of α (intercept) and γ (slope) and their uncertainties ($\Delta\alpha$ and $\Delta\gamma$) were estimated with the non-parametric technique of bootstrapping.

4. Results

4.1 Catalogs of landslides and rainfall events at regional scale

Using the Thiessen polygon technique, six rainfall regions were identified at the regional scale: three in the South (Japuiba, Angra dos Reis, and Jacuecanga – JAJ) and three in the North (Mambucada, Bracui, and Serra d'Agua – MBS). Then, the

	Extent	RE	LE	D (h)			E (mm)		
	(km²)			Min	Max	Mean	Min	Max	Mean
Municipality	825	484	1640	24	624	101,3	2	542,9	111,9
South: JAJ	374	357	1276	24	624	102,2	2	542,9	112,8
North: MBS	451	127	364	24	432	97,9	2	400,8	108,7
Urban Center	6,5	129	526	24	528	108,4	2	540,5	114,9

Table 1.
Statistics of rainfall events (RE), landslide events (LE), duration (D) and accumulation of rainfall (E) that initiated landslides at the regional scale (municipality, north, south) and local scale of the urban center of Angra dos Reis.

1640 landslides recorded were associated with one of the 1434 rainfall events compiled (≥2 mm). In that regards, only 33% (484 out of 1434) of these rainfall events have triggered landslides (**Table 1**).

A North–South disparity appears in the number of rainfall events recorded in the North (127) and South (357) of the municipality (**Table 1**). The number of recorded landslides is significantly higher in the South (1274) compared to the North (364). Therefore, 78% of the landslides occurred in the South (JAJ) of the region (**Figure 1**), where the majority of the population and urban areas are concentrated. On the other hand, few landslides were recorded in the North (MBS) and in the vegetated areas of the Bocaína Plateau (**Figure 1**).

Regarding the inter-annual variability of rainfall events triggering landslides for the 6-years period analyzed indicates that landslides occur every year. On an intra-annual basis, the average number of triggering events and the ratios of triggering

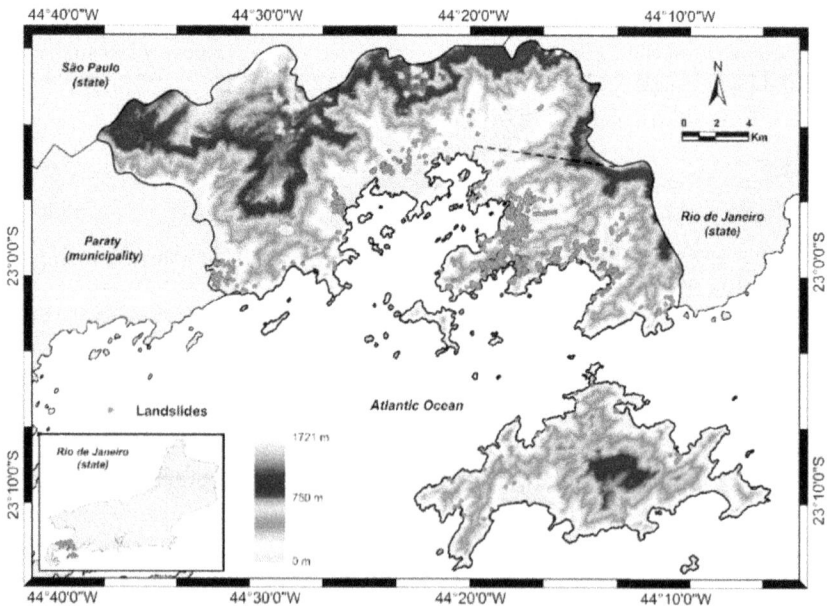

Figure 1.
Map of the study area (825 km²), Municipality of Angra dos Reis in the state of Rio de Janeiro, Brazil. Red dots show location of the 1640 landslides recorded and the black dashed line represents the limit between north (Mambucada, Bracui, and Serra d'Agua – MBS) and south (Japuida, Angra dos Reis, and Jacuecanga – JAJ) regions.

versus non-triggering events vary primarily with the seasonality. Indeed, the average numbers and ratios are 9 and 40% in the wet season (January to April), 5 and 25% in the dry season (May to August), and 8 and 32% in the transition season (September to November).

The 484 rainfall events that likely triggered landslides lasted approximately 4 days (101 hours; **Table 1**), with a minimum and maximum duration of 24 and 624 hours (26 days), and initiated an average of three landslides. Specifically, 45% (219 out of 484) of the triggering events initiated a single slide, 75% (365 out of 484) triggered three or fewer slides, and only 8% (37 out of 484) generated ten or more failures, with a maximum of 38 landslides per episode. In this regard, two rainfall events recorded in the region of Angra dos Reis triggered exactly 38 landslides. The first one started on December 27, 2012 and ended on January 4, 2013, accumulating 540.5 mm of rain in nine days, while the second one started on January 9, 2013 and ended on January 22, 2013 after dumping 372.9 mm in 14 days. While the first of the two rainfall events had accumulated only 46.5 mm in the previous 30 days, the second had accumulated 565 mm of antecedent rainfall over 30 days. Therefore, the soils of the region had received 937.9 mm in 44 days as of January 22, 2013. Angra dos Reis was specifically the region the most affected by landslides during the study period, accounting for 39% (638 out of 1640) of all landslides recorded. This corresponds to an average of four landslides per triggering rainfall event, which is slightly higher than the regional average of three at the municipal level.

On a seasonal basis, 69% of the landslides (1130 of 1640) were initiated during the wet season. This significantly outweighs the amount of landslides that were initiated during the dry season and the transition season, both of which accounted for approximately 15%. On a monthly basis, January had the most landslides initiated, followed by March, April and December in relatively equal proportions. This average of 69 landslides per month in January is almost twice as the amount recorded in the other wet months (December, February, March, and April), which average 33, and five times more than in drier months (May to October), which average 13. Finally, the month of May represents the least likely period for landslide occurrence with only six landslides recorded on average. This data is nevertheless significant and indicates that landslides can occur in any month of the year, despite less precipitation in the winter period (May to August).

With respect to slope steepness, 71% of the cases occurred on gentle slopes (0 to 20°), 27% on slopes between 20 and 35° and only 2% on steep slopes (>35°). Regarding slope orientation, more than a third of the landslides (34%) were initiated on south facing slopes, i.e. facing the prevailing winds, while 24% were initiated on north facing slopes. The remaining cases were associated with west (18%) and east (16%) facing slopes and relatively flat terrain (8%). Finally, 58% of the landslides were triggered in urban areas compared to 27% in forested areas and 8% in pastures.

4.2 Catalogs of landslides and rainfall events at local scale

At the urban center scale of Angra dos Reis, 526 landslides were linked to one of 234 rainfall events compiled. The geographic distribution of landslides is relatively heterogeneous despite a fairly large clustering (155 cases; 30%) in the colluviums of the Sapinhatuba I and Monte Castelo neighborhoods in the east-central part of the urban center (**Figure 2**). The landslides were initiated during 129 of the 234 (55%) rainfall events compiled (**Table 1**). These 129 rainfall events lasted a little more than 4 days, or 108 hours, with a minimum and maximum duration of 24 and 528 hours (22 days). The minimum and maximum number of landslides initiated by

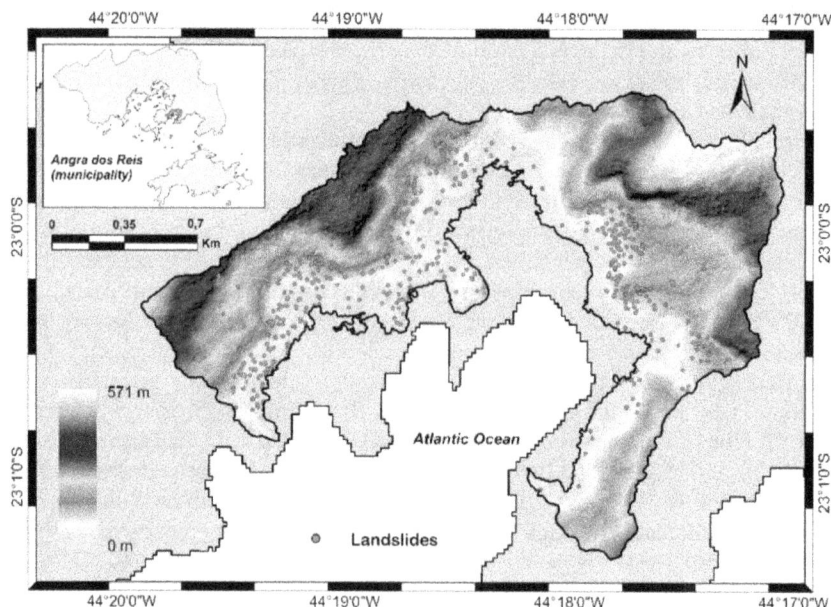

Figure 2.
Map of the urban center of Angra dos Reis. Red dots show location of the 526 landslides recorded over an area of 6.5 km².

these episodes are 1 and 38, for an average of four landslides per triggering rainfall event. More specifically, 36% (47 of 131) of the triggering rainfall events initiated a single slide, 63% (83 of 131) initiated three or fewer slides, and only 9% (12 of 131) generated ten or more ruptures.

On a seasonal basis, two-thirds of the 526 slides (66%) were initiated during the wet season. Indeed, the average number of rainfall triggering events and the ratios of triggering versus non-triggering events are 2 and 64% in the wet season (January to April), 1 and 37% in the dry season (May to August), and 2 and 59% in the transition season (September to November). On a monthly basis, January was the most significant month with an average of 25 landslides compared to only seven for all other months. 42% of the landslides occurred on gentle slopes (0 to 20°), 39% on slopes between 20 and 35° and 19% on steep slopes >35°. In this respect, more landslides were recorded on steep slopes at the local scale compared to the regional scale. The slopes facing south were the most affected with 53% of the landslides, compared to 21% on slopes facing north, 13% on the east and 14% on the west facing slopes (**Figure 2**). Finally, a high proportion of landslides (76%) occurred in urban areas, while only 22% occurred in forest and pasture areas.

4.3 Definition of the rainfall thresholds for landslide events

As we mentioned in the methodological section, several ED thresholds were defined at the regional and local scales based on the overall database, but also with different subsets for the seasonality (wet and dry seasons), and the southern (JAJ) and northern (MBS) parts of the study area.

Figure 3 shows, in log–log coordinates, the distribution of rainfall conditions (D, E) that have resulted in landslides at both scales; the municipality of Angra

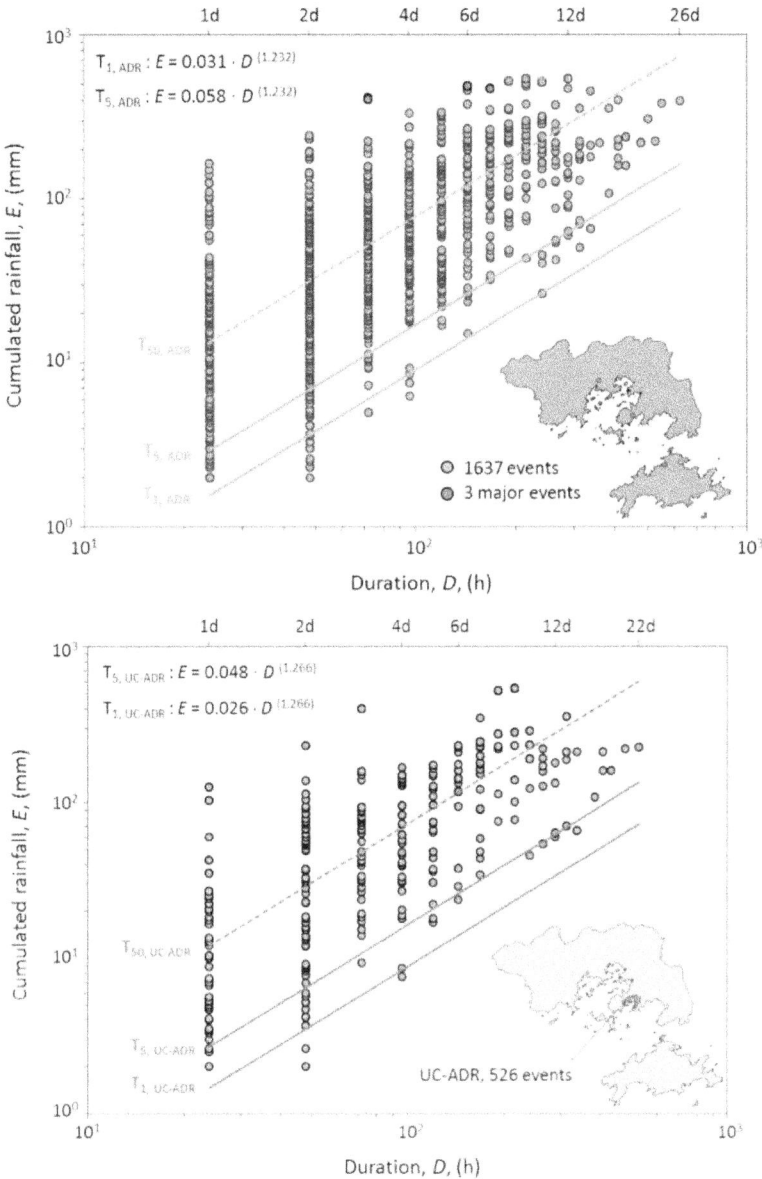

Figure 3.
Rainfall duration D (x-axis) and cumulated event rainfall E (y-axis) conditions that have resulted in landslides at the regional scale of the municipality of Angra dos Reis (upper panel) and at the local scale of the urban center (lower panel). 1% and 5% ED power law thresholds are shown with their equations (solid lines) as well as 50% (dashed line) for reference. Inset shows the study area and the localization of the three major deadly events (red dots) at the regional scale.

dos Reis (1640 landslides) and the urban center (526 landslides), with the 1% and 5% ED thresholds curves. The urban center shows a quite similar 1% ED threshold ($T_{1\,UC\text{-}ADR}$) and 5% ED threshold ($T_{5\,UC\text{-}ADR}$) to those calculated at the regional scale ($T_{1\,ADR}$ and $T_{5\,ADR}$). See **Table 1** for more details about the scale and intercept parameters.

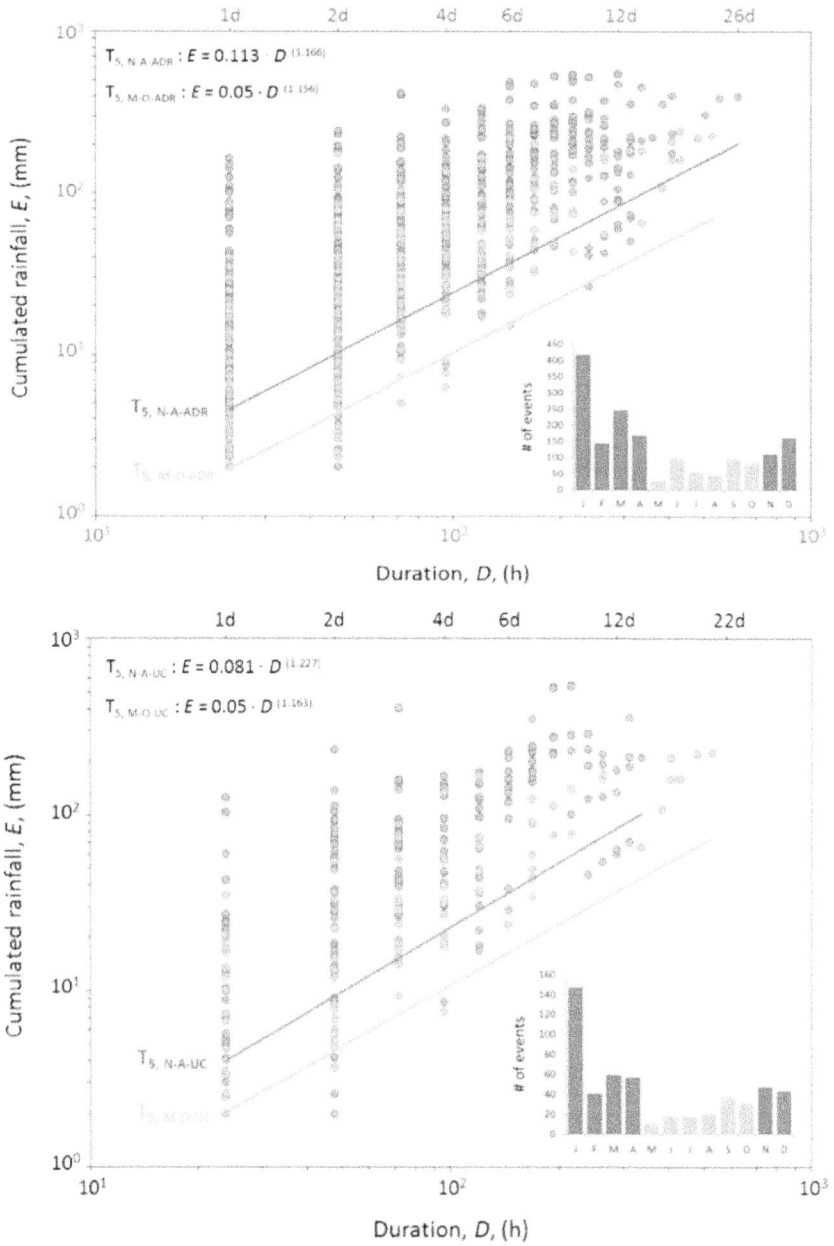

Figure 4.
Rainfall duration D (x-axis) and cumulated event rainfall E (y-axis) conditions that have resulted in landslides in the period May–October (dry season; blue dots) and for the period November–April (wet season; gray dots). Colored lines are the 5% power law thresholds and insets show the distribution of recorded landslides on a monthly basis. Upper panel shows the dataset at the regional scale and lower panel shows the dataset at the local scale of urban center.

The ED rainfall thresholds shown in **Figure 4** indicate lower 5% thresholds for the extended dry season ($T_{5\ M\text{-}O\text{-}ADR}$ and $T_{5\text{-}M\text{-}O\text{-}UC}$) by comparison to the wet season ($T_{5\text{-}N\text{-}A\text{-}ADR}$ and $T_{5\ N\text{-}A\text{-}UC}$) at both scales. Indeed, the insets clearly indicate the high monthly variability in occurrence of landslides recorded, showing obvious differences between the dry and wet season. In this case, the 5% ED thresholds were quite similar for the municipality and urban center, particularly for the dry season and with a very small difference for the wet season (**Table 1**). However, the rainfall events triggering landslides during the wet season appears of shorter duration at the local scale (24 to 336 hours) compared to the regional scale (24 to 624 hours). No difference was reported for the dry season.

Label	Area	RE	LE	Threshold	Range (h)	Uncertainty
$T_{1, ADR}$	R	484	1640	$E = 0.031xD^{1.232}$	24-624	$\Delta\alpha = 0.13\ \Delta\gamma = 0,002$
$T_{5, ADR}$	R	484	1640	$E = 0.058xD^{1.232}$	24-624	$\Delta\alpha = 0.18,\ \Delta\gamma = 0.002$
$T_{1, UC-ADR}$	L	129	526	$E = 0.026xD^{1.266}$	24-528	$\Delta\alpha = 5.32\ \Delta\gamma = 0.04$
$T_{5, UC-ADR}$	L	129	526	$E = 0.048xD^{1.266}$	24-528	$\Delta\alpha = 0.7\ \Delta\gamma = 0.03$
$T_{5, N-A-ADR}$	R-Wet	1240	1240	$E = 0.113xD^{1.166}$	24-624	$\Delta\alpha = 0.19\ \Delta\gamma = 0.02$
$T_{5, M-O-ADR}$	R-Dry	400	400	$E = 0.05xD^{1.256}$	24-528	$\Delta\alpha = 0.20\ \Delta\gamma = 0.08$
$T_{5, N-A-UC}$	L-Wet	393	393	$E = 0.081xD^{1.227}$	24-336	$\Delta\alpha = 0.22\ \Delta\gamma = 0.03$
$T_{5, M-O-UC}$	L-Dry	133	133	$E = 0.05xD^{1.163}$	24-528	$\Delta\alpha = 1.2\ \Delta\gamma = 0.02$
$T_{5, JAJ-ADR}$	R-South	357	1276	$E = 0.056xD^{1.228}$	24-624	$\Delta\alpha = 0.12\ \Delta\gamma = 0.02$
$T_{5, MBS-ADR}$	R-North	451	364	$E = 0.066xD^{1.252}$	24-432	$\Delta\alpha = 0.6\ \Delta\gamma = 0.04$

Label: label of the thresholds defined in this study. Area: regional scale (R) of the municipality or local scale (L) of the urban center. RE: number of rainfall events. LE: number of landslide events. Threshold: D rainfall duration, in hours; E, cumulated event rainfall, in mm. Range: range of the validity for the threshold. Uncertainty: associated with the intercept α and the slope γ of the threshold model based on a power law.

Table 2.
Rainfall ED thresholds for the possible initiation of landslides in Angra dos Reis.

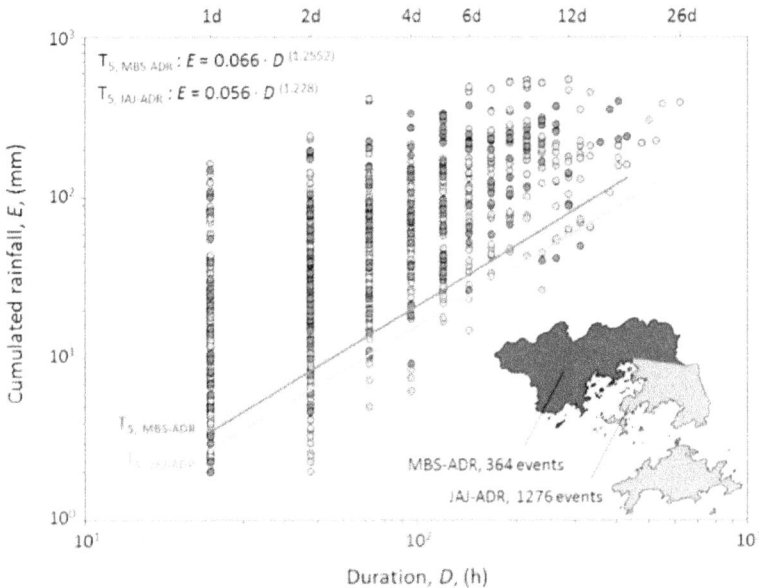

Figure 5.
Rainfall duration D (x-axis) and cumulated event rainfall E (y-axis) conditions that have resulted in landslides in the southern region (JAJ; yellow dots) and the northern region (MBS; red dots). Colored lines are the 5% power law thresholds and inset shows the geographical area covers by both regions as well as the number of landslides recorded.

Finally, the **Figure 5** shows the similar tendency of the 5% thresholds for the southern and northern regions, although the latter shows a slightly higher value. However, it is worth mentioning the significant difference between the two regions regarding the number of recorded landslides; 1276 in the southern part and 364 in the northern part of the municipality. A significant appears also in the duration of rainfall events triggering landslides, which is limited to 432 hours in the north by comparison to 624 hours in the south (**Table 2** and **Figure 5**).

5. Discussion

5.1 Landslide hazard and risk in Brazil

Several factors of a geological (volcanic activity, earthquake, lithologic faults and discontinuities, etc.), meteorological (precipitation, temperature), and human (land use) nature influence slope stability [36]. Therefore, the dynamic relationships in time and space between these different factors greatly complicate the objective assessment of landslide susceptibility and probability of occurrence for a given region and time period [37].

In the humid tropics, the majority of precipitation and its extremes are concentrated during the summer period. To this end, the warm and humid climate of southeastern Brazil favors the chemical alteration of rock and the development of saprolites, especially in hilly and mountainous environments. During major rainfall events, these are reworked by mass movements such as debris flows, superficial slides, rotational, and deep-seated landslides [38]. It is therefore not surprising that in the past several catastrophic events occurred [39–41], for example and among others in 1967 [42], 1988 and 1996 [43–45], and more recently in 2008, 2010 and 2011 [46–48]. The rapid growth of urbanization in the last decades with an improper land use [38] is certainly responsible, at least partially, for the tragic outcome of these recent disasters.

In this context, several approaches have subsequently been used to assess landslide risk: landslide susceptibility zonation using GIS-based fuzzy logic [49], electric and electromagnetic methods with geotechnical soundings [50], structural geology and kinematic analysis with stereographic projections [51, 52], analysis of morphological parameters (drainage efficiency index, slope geometry, slope angles, etc.) [53], laboratory and fields observations [54], and modeling with SHALSTAB (shallow landsliding stability model), SINMAP (stability index mapping), GEO-SLOPE (slope stability analysis), and TRIGRS (transient rainfall infiltration) software's [55–57]. Despite the limited effectiveness of these modeling procedures and the interest in mapping and vulnerability of populations to landslides, e.g. [58–60], there is still little work on the determination of rainfall thresholds favorable to landslide occurrence. Indeed, the few authors who have focused on defining rainfall thresholds in southeastern Brazil and the Serra do Mar at the regional scale, are rather thresholds based on total precipitation during major events [19, 61].

5.2 Significance of ED rainfall thresholds for hazard and risk assessment

The ED thresholds presented in this study at regional and local scales highlight the statistical dependency of the cumulated rainfall E to the rainfall duration D. In that regards, they represent appropriate rainfall thresholds for the possible occurrence of 1640 landslides in the municipality of Angra dos Reis and 526 landslides at the local scale of the urban center over a 6-year period. However, because these thresholds result of statistical modeling applied to an empirical dataset, uncertainties have been quantified using a bootstrap approach.

Figure 4 indicates a significant difference in seasonal thresholds for all duration analyzed. Surprisingly, less cumulated rainfall appear required to initiate landslides during the dry season (May–October) by comparison to the wet summer season (November–April). This result was not expected considering that usually the ante-cedent rainfall conditions of the wet season, and the resulting increased moisture in the soil, reduce the amount of event rainfall required to triggering landslides [30]. Unfortunately, the absence of details about the landslides recorded does not allow a better discrimination in landslide classification or typology regarding their temporal occurrence. However, the significant difference between the northern and southern regions in number of landslides triggered by rainfall events (**Figure 5**), attests for the likely influence of other environmental factors such as slope aspect, land cover type, lithological types, etc. Slope steepness does not appear to be a very important factor given the amount of landslides recorded on gentle slopes. On the other hand, even considering the size of the dataset analyzed (484 rainfall events resulting in 1640 landslides), we acknowledge that further studies are required to better understand the role of land use, land cover types, urbanization, and human induced changes that may affect the amount of rain necessary to trigger landslides locally and regionally. Finally, as mentioned by [62], the bootstrapping technique may result in optimistic estimates of the uncertainties in the thresholds determined, to which we suggest conducting similar analyses over longer time series.

In the Brazilian municipality of Angra dos Reis, heavy summer rainfall has historically triggered several landslides, causing a significant number of victims and considerable damage. The risk posed by this geomorphic hazard is particularly high due to the fact that almost 60% of the population lives in slope areas [16] and more than 25% live in areas considered to be at high risk of landslides [17]. In that regards, our data indicate that 58% of the 1640 landslides recorded occurred in urban areas, 71% on gentle slope, and 34% on south facing slopes, reflecting the exposure and risk to the population. Therefore, the interest of the municipal authorities of Angra dos Reis in establishing rainfall thresholds (i.e. **Table 2**) should allow a better anticipation of spatial and temporal occurrence of the phenomenon. The thresholds in this study could ultimately be integrated into a landslide monitor-ing and warning system and serve as a necessary component of hazard assessment. This is particularly pertinent considering that since the beginning of research on rain-triggered landslides in Brazil, no suitable thresholds have been proposed for the Angra dos Reis territory.

In fact, the thresholds of Guidicini and Iwasa [18] were established for the whole Serra do Mar, whose area is four times larger than that of Angra dos Reis, while the one developed by Soares [19] is not representative of the triggering conditions due to the use of an inadequate database. The thresholds proposed and to come in the near future according to the characteristics of the territory and from recent and reliable data could be an effective tool for landslide risk management.

6. Conclusion

In the mountainous and tropical environment of the municipality of Angra dos Reis in Brazil, the high frequency of intense rainfall generates several landslides that recurrently interfere with human activities and infrastructures. Lithology, land use and vegetation cover are biophysical parameters that remain to be explored in relation to the spatial and temporal dynamics of landslides, especially in a context of climate change and increasing urbanization.

The establishment of quantitative rainfall thresholds that, when reached or exceeded, are likely to trigger landslides (e.g. **Figures 3–5**) therefore appears to be a valid approach for risk management. The thresholds reported in this study could provide a relevant management tool for municipal authorities. Moreover, the establishment of thresholds based on the duration of rainfall events (ED) should be regarded as a research axis whose development is essential in risk management, particularly in order to set up a landslide monitoring and warning system. The detailed study of rainfall conditions that led to the initiation of 1640 landslides in the municipality of Angra dos Reis and 526 landslides in the urban center revealed that very small amounts of water accumulated over periods of up to 26 days are sufficient to initiate landslides (**Table 2**). These precipitations represent barely 1 to 4% of the annual average rainfall depending on the duration of the events considered. The rainfall limits appear low when compared to some of the thresholds proposed in the literature for Brazil and other tropical regions.

Acknowledgements

Special thanks to the municipal authorities of Angra dos Reis for providing us the data on the occurrence of landslides as well as the data from the rainfall stations in their territory. Finally, thanks also to the Scientific Council of the Brazilian Government for granting research funding in the context of this Canada-Brazil cooperation, particularly with the department of Geography of the State University of Rio de Janeiro.

Author details

Daniel Germain[1*], Sébastien Roy[1] and Antonio Jose Teixera Guerra[2]

1 Department of Geography, Université du Québec à Montréal, Montréal, Canada

2 Department of Geography, Federal University of Rio de Janeiro, Rio de Janeiro, Brazil

*Address all correspondence to: germain.daniel@uqam.ca

IntechOpen

References

[1] Corominas J, Moya J. Reconstructing recent landslide activity in relation to rainfall in the llobregat river basin, Eastern Pyrenees, Spain. Geomorphology. 1999;30:79-93

[2] Cepeda J, Höeg K, Nadim F. Landslide-triggering rainfall thresholds: a conceptual framework. Quarterly Journal of Engineering Geology and Hydrogeology. 2010;43:69-84

[3] Iverson R. Landslide triggering by rain infiltration. Water Resources Research. 2000;36:1897-1910

[4] Duc DM. Rainfall-triggered large landslides on 15 December 2005 in Van Canh District, Binh Dinh Province, Vietnam. Landslides. 2013;10:219-230

[5] Fernandes NF, Guimarães RF, Gomes RAT, Vieira BC, Montgomery DR, Greenberg H. Topographic controls of landslides in Rio de Janeiro: field evidence and modeling. Catena. 2004;55:163-181

[6] Berti M, Martina MLV, Franceschini S, Pignone S, Simoni A, Pizziolo M. Probabilistic rainfall thresholds for landslide occurrence using a bayesian approach. Journal of Geophysical Research. 2012;117:20

[7] Ribeiro MF, da Costa VC, Neto NM, de Freitas, MAV. An analysis of monthly rainfall and its relationship to the occurrence of mass movement and flooding in Pedra Branca Massif in the city of Rio de Janeiro, Brazil. Geographical Research. 2013;51:398-411

[8] SafeLand. 2012. Landslide triggering mechanisms in Europe – overview and state of the art. 7th framework program, 378 p.

[9] Martelloni G, Segoni S, Fanti R, Catani F. Rainfall thresholds for the

[10] Caine N. The rainfall intensity-duration control of shallow landslides and debris flows. Geografiska Annaler: Series A, Physical Geography. 1980;62:23-27

[11] Dahal RK, Hasegawa S. Representative rainfall thresholds for landslides in the Nepal Himalaya. Geomorphology. 2008;100:429-443

[12] Aleotti P. A warning system for rainfall-induced shallow failure. Engineering Geology. 2004;73:247-265

[13] Talebi A, Nafarzadegan AR, Malekinezhad H. A review of empirical and physically based modeling of rainfall triggered landslides. Physical Geography Research Quaterly. 2010;70:45-64

[14] Jakob M, Weatherly H. An hydroclimatic threshold for landslide initiation on the north shore mountains of Vancouver, British Columbia. Geomorphology. 2003;54:137-156

[15] Vieira BC, Fernandes NF, Filho OA. Shallow landslide prediction in the Serra do Mar, São Paulo, Brazil. Natural Hazards and Earth System Sciences. 2003;10:1829-1837

[16] Bortoloti M. 2010. Trágico, absurdo, previsível [Internet]. 2010. Available from: http://veja.abril.com.br/130110/tragico-absurdo-previsivel-p-054.shtml [Accessed: 2014-03-22]

[17] Geological Survey of Brazil. Riscos geológicos [Internet]. 2014. Available from: http://www.cprm.gov.br/publique/cgi/cgilua.exe/sys/start.htm?sid=38 [Accessed: 2014-03-22]

[18] Guidicini G, Iwasa OY. Essai de corrélation entre la pluviosité et les

glissements de terrain sous climat tropical humide. Bulletin de l'Association Internationale de Géologie de l'Ingénieur. 1977;16:13-20

[19] Soares EP. Caracterização da precipitação na região de angra dos reis e a sua relação com a occorréncia de deslizamentos de encostas [thesis]. Rio de Janeiro, Universidade Federal do Rio de Janeiro; 2006

[20] Coppetec. Mapeamento de áreas de riscos, frente aos deslizamentos de encostas no município de Angra dos Reis. Coppe/Universidade Federal do Rio de Janeiro, Reports 2-4; 2011

[21] de Almeida, FDM, Carneiro CDR. Origem e evolução da serra do mar. Revista Brasileira de Geociências. 1998;28:135-150

[22] Ferrari A, Mansur K. Ponto de interesse geológico: Serra do Mar. Rio de Janeiro: projeto caminhos geológicos; 2012

[23] Geological Survey of Brazil. Geologia da folha Volta Redonda sf.23-z-a-v. Contracto cprm-uerj no. 057/pr/05. Brasília: programa de geologia do Brasil; 2007. 148 p.

[24] Soares FS, Francisco CN, Senna MCA. Distribuição espaço-temporal da precipitação na região hidrografica da baía da Ilha Grande – Rio de Janeiro. Revista Brasileira de Meteorologia. 2014;29;125-138

[25] de Souza, FT, Ebecken NNF. A data based model to predict landslide induced by rainfall in Rio de Janeiro City. Geotechnical and Geological Engineering. 2012;30:85-94

[26] Mazza BC. Inventário de movimentos de massa gravitacionais na Serra do Mar no municipio de Angra dos Reis, Rio de Janeiro [thesis]. Rio de Janeiro, Universidade Federal Rural do Rio de Janeiro; 2007

[27] Guerra AJT, Bezerra, JFR, Jorge MDCO, Fullen MA. The geomorphology of Angra dos Reis and paraty municipalities, southern Rio de Janeiro State». Revista Geonorte. 2013;8:1-21

[28] Pocidonio EAL, da Silva TM. Nature as attraction and repulsion in the city of Angra dos Reis, Rio de Janeiro State. Geo Uerj. 2011;2:422-446

[29] Instituto Brasileiro de Geografia e Estatística. 2014. Online Database [Internet]. 2014. Available from: http://www.ibge.gov.br/home/ [Accessed: 2014-01-13]

[30] Zêzere JL, Trigo RM, Trigo IF. Shallow and deep landslides induced by rainfall in the Lisbon region (Portugal): assessment of relationships with the North Atlantic Oscillation. Natural Hazards and Earth System Sciences. 2005;5:331-344

[31] Brunetti MT, Peruccacci S, Rossi M, Luciani S, Valigi D, Guzzetti F. Rainfall thresholds for the possible occurrence of landslides in Italy. Natural Hazards and Earth System Sciences. 2010;10:447-458

[32] Environmental Systems Research Institute. Arcgis, Redland: ESRI; 2013

[33] Thiessen AJ, Alter JC. Precipitation averages for larges areas. Monthly Weather Review. 1911;39:1082-1084

[34] Guzzetti F, Peruccacci S, Rossi M, Stark CP. Rainfall thresholds for the initiation of landslides in central and southern Europe. Meteorology and Atmospheric Physics. 2007;98:239-267

[35] Peruccacci S, Brunetti MT, Luciani S, Vennari C, Guzzetti F. Lithological and seasonal control on rainfall thresholds for the possible initiation of landslides in central Italy. Geomorphology. 2012;139-140:79-90

[36] Gariano SL, Guzzetti F. Landslides in a changing climate. Earth-Science Reviews. 2016;162 :227-252

[37] Borgomeo E, Hebditch KV, Whittaker AC, Lonergan L. Characterising the spatial distribution, frequency and geomorphic controls on landslide occurrence, Molise, Italy. Geomorphology. 2014;226:148-161

[38] Alheiros MM, Filho OA. Landslides and coastal erosion hazards in Brazil. International Geology Review. 1997;39:756-763

[39] Avila A, Justino F, Wilson A, Bromwich D, Amorim M. Recent precipitation trends, flash floods and landslides in southern Brazil. Environmental Research Letters. 2016;11:114029

[40] Guerra A. Catastrophic events in Petropolis City (Rio de Janeiro State), between 1940 and 1990. Geojournal. 1995;37:349-254

[41] Kabiyama M, Michel GP, Engster EC, Paixao MA. Historical analyses of debris flow disaster occurrences and their scientific investigation in Brazil. Labor & Engenho. 2015;9:76-89

[42] Dias HC, Dias VC, Vieira BC. 2016. Landslides and morphological characterization in the Serra do Mar, Brazil. In: Aversa et al. editors. Landslides and Engineered Slopes. Experience, Theory and Practice. Associazione Geotecnica Italiana; 2016. p. 831-836

[43] Coelho-Netto AL. Produção de sedimentos em bacias fluviais florestadas do maciço da Tijuca: respostas aos eventos extremos de fevereiro de 1996. Anais do ii Encontro Nacional de Engenharia de Sedimentos. 1996;1:209-217

[44] Coelho-Netto Al. Catastrophic landscape evolution in a humid region

SE Brasil: inheritances from tectonic, climatic and land use induced changes. Geografia Fisica e Dinamica Quaternaria. 1999;3:21-48

[45] Lacerda WA. Stability of natural slopes along the tropical coast of Brazil. In : Almeida MSS. editor. Symposium on recent developments in soil and pavement mechanics, Balkema, Rotterdam; 1997. p. 17-40

[46] Assis Dias MC, Saito SM, Alvala RC, Stenner C, Pinho G, Nobre CA, Fonseca MR, Santos C, Amadeu P, Silva D, Lima CO, Ribeiro J, Nascimento F, Correra CO. Estimation of exposed population to landslides and floods risk areas in Brazil, on an intra-urban scale. International Journal of Disaster Risk Management. 2018;31:449-459

[47] Graeff O, Guerra A, Jorge MC. Floods and landslides in Brazil. A case study of the 2011 event. Geography Review. 2012;September:38-41

[48] Coelho-Netto AL, Sato AM, Avelar AS, Vianna LGG, Araujo IS, Ferreira DLC, Lima PH, Silva APA, Silva RP. January 2011: The extreme landslide disaster in Brazil. In: Margottini C. et al. editors. Landslide Science and Practice –The second World Landslide Forum; 2013

[49] Bortoloti FD, Castro Junior RM, Araujo LC, de Morais GB. Preliminary landslide susceptibility zonation using gis-based fuzzy logic in Victoria, Brazil. Environmental Earth Sciences. 2015;74:2125-2141

[50] Bortolozi CA, Motta MFB, de Andrade MCM, Lavalle LVA, Mendes RM, Simoes SJC, Mendes TSG, Pampuch LA. Combined analysis of electric and electromagnetic methods with geotechnical soundings as soil characterization as applied to a landslide study in Campos do Jordao City, Brazil. Journal of Applied Geophysics. 2019;161:1-14

[51] Cerri RI, Reis AGV, Gramani MF, Giordano LC, Zaine JE. Landslides zonation hazard: relation between geological structures and landslides occurrence in hilly tropical regiona of Brazil. Anaia da Academia Brasileria de Ciencias. 2017;89:2609-2623

[52] Cerri RI, Reis AGV, Gramani MF, Rosolen V, Luvizotto GL, Giordano LC, Gabelini BM. Assessment of landslide occurrences in Serra do Mar Mountain Range using kinematic analyses. Environmental Earth Sciences. 2018;77:325

[53] Coelho-Netto AL, Avelar AS, Fernandes MC, Lacerda WA. Landslide susceptibility in a mountainous geoecosystem, Tijuca Massif, Rio de Janeiro: the role of morphometric subdivision of the terrain. Geomorphology. 2007;87:120-131

[54] Lacerda WA. Landslide initiation in saprolite and colluvium in southern Brazil : field and laboratory observations. Geomorpohlogy. 2007;87 :104-119

[55] Michel GP, Kobiyama M, Goerth RF. Comparative analysis of SHALSTAB and SINMAP for landslide susceptibility mapping in the Cunha river basin, southern Brazil. Journal of Soils and Sediments. 2014;14:1266-1277

[56] Mendes RM, de Nadrade MRM, Tomasella J, de Moraes MAE, Scofield GB. Understanding shallow landslides in Campos do Jordao Municipality – Brazil : disentangling the anthropic effects from natural causes in the disaster of 2000. Natural Hazards and Earth System Science. 2018;18:15-30

[57] Vieira BC, Fernandes NF, Filho OA, Martins TD, Montgomery DR. Assessing shallow landslide hazards using the TRIGRS and SHALSTAB models, Serra do Mar, Brazil. Environmental Earth Sciences. 2018;77:260

[58] Listo FLR, Vieira BC. Mapping risk and susceptibility of shallow-landslide in the city of Sao Paulo, Brazil. Geomorphology. 2012;169-170:30-44

[59] Debortoli NS, Camarinha PIV, Marengo JA, Rodrigues RR. An index of Brazil's vulnerability to expected increases in natural flash flooding and landslide disasters in the context of climate change. Natural Hazards. 2017;86:557-582

[60] Batista JAN, Julien PY. Remotely sensed survey of landslide clusters : case study of Itaoca, Brazil. Journal of South American Earth Sciences. 2019;92:145-150

[61] Almeida MCJ, Nazakawa A, Tatizana C. Análise de correlação entre chuvas e escorregamentos no Município de Petrópolis, Rio de Janeiro. In: Proceedings of the 7° Congresso Brasileiro de Geologia de Engenharia; 12-16 September 1993; São Paulo. ABGE; 1993. p. 129-133

[62] Efron B, Tibshirani RJ. An Introduction to the Bootstrap. Chapman and Hall; 1994

www.ingramcontent.com/pod-product-compliance
Lightning Source LLC
Chambersburg PA
CBHW081534190326
41458CB00015B/5548